MONOGRAPHS
ON THE PHYSICS AND CHEMISTRY OF MATERIALS

General Editors

H. FRÖHLICH

P. B. HIRSCH

N. F. MOTT

ELECTRONS IN LIQUID AMMONIA

BY

J. C. THOMPSON

CLARENDON PRESS · OXFORD

1976

Oxford University Press, Walton Street, Oxford OX2 6DP

GLASGOW NEW YORK TORONTO MELBOURNE WELLINGTON
CAPE TOWN IBADAN NAIROBI DAR ES SALAAM LUSAKA ADDIS ABABA
DELHI BOMBAY CALCUTTA MADRAS KARACHI
KUALA LUMPUR SINGAPORE HONG KONG TOKYO

ISBN 0 19 851343 7

FILMSET IN N. IRELAND AT THE UNIVERSITIES PRESS, BELFAST
PRINTED IN GREAT BRITAIN
BY J. W. ARROWSMITH LTD., BRISTOL

PREFACE

SOLUTIONS of metals in liquid ammonia have been known for over a century and studies of their properties have resulted in hundreds of papers. This enduring interest is sustained by the extraordinarily rich variety of properties which is observed as the concentration of metal is changed. Electrical conductivities range from 10^{-8} to $10^4 \Omega^{-1} \, \text{cm}^{-1}$; spin resonance lines are the narrowest known; densities are lower only for the cryogenic fluids; and the blue color characteristic of the dilute solutions has become the hallmark of the lightest anion—the solvated electron. Whether one's interests are chemical or physical, whether one studies kinetics, reactions, electrolytes, critical phenomena, metals, insulators, or the metal–insulator transition, there are even now experiments to be done, models to be constructed and understanding to be gained. Indeed, developments at the 4th Colloque Weyl have added to our knowledge of the metal–nonmetal transition and the solid hexammines.

I have not attempted to cover all the available studies of M–NH$_3$ solutions, rather limiting myself to those topics which come under the headings of chemical physics and physical chemistry.† In the chapters which follow I have tried to bridge the gap between readers with chemical and those with physical background, making the arguments intelligible to both (at the risk of explaining the obvious to some). I hope to find an audience in both camps. Section 1.4 contains a more detailed outline of the contents.

There are disadvantages to this wealth as well, for a single individual cannot adequately review all of the available information and my prejudices and ignorance have doubtless led me to slight work deserving of high praise. For this I apologize.

I must not, however, slight those who have provided me with guidance and support during the years I have studied M–NH$_3$ solutions. The latter has come from the U.S. Office of Naval Research, monitored by Jack Soules, from the U.S. National Science Foundation, monitored by Langdon Crane and Rolf

---

† One can find an introduction to more chemical topics in the series *Chemistry of non-aqueous solvents*, edited by J. J. Lagowski.

Sinclair, and from the R. A. Welch Foundation of Texas, directed by W. O. Milligan. These agencies and individuals have sustained me not only by their funds but also by their trust and encouragement, which I deeply appreciate.

Guidance has come from many students, friends, and colleagues, but two individuals have contributed more than the rest: Morrel Cohen and Nevill Mott. Morrel and I came together one weekend in 1967. I was possessed by a mountain of data while he, as always, had the experience and insight to help me organize and begin to understand what it was that I was about. Sir Nevill I met at 'his' conference in 1968. Since then in conversation and correspondence he has shared his extraordinary intuition with me. In their separate ways and with repeated kindness they have helped me along the way. They have my eternal gratitude.

Most authors acknowledge the sustenance, encouragement, and patience of their wives, and this debt I certainly owe to my wife, Carol May. But I owe her infinitely more, for it was she who introduced me to the existence and excitement of metal–ammonia solutions and who, with skill and tact, taught me the chemistry I needed to read and to understand the vast literature behind this book. And, finally, it was she who strove to reduce my bad English and bad science. For all this, and more, I owe her everything. That this book exists at all is a tribute to her.

Austin, Texas   J.C.T.
April, 1976

ACKNOWLEDGEMENTS

The publishers of the following journals and books and the authors are due acknowledgement for their permission for reproduction of the figures specified: Academic Press, from *Chemistry of non-aqueous solvents*, Vol. 2 (Fig. 5.10); *Adv. Phys.* (Fig. 2.24, 2.25, 3.33); International Union of Pure and Applied Chemistry, from *Metal-ammonia solutions* (Fig. 3.3, 3.36, 4.3, 7.2); *J. Chim. phys.* (Fig. 3.32); *J. chem. Engng. Data* (Copyright Am. Chem. Soc.) (Fig. 3.22, 4.4); *J. chem. Phys.* (Fig. 2.6, 2.7, 2.10, 2.11, 2.12, 2.13, 2.14, 2.15, 2.20, 2.28, 3.10, 3.14, 3.20, 3.21, 3.26, 3.34, 3.37, 3.38, 4.1, 4.6, 5.1, 5.6, 7.3, 7.9); *J. phys. Chem.* (Copyright Am. Chem. Soc.) (Fig. 3.1, 3.17, 3.19, 5.7, 6.6, 8.2); *J. Solid State Chem.* (Fig. 7.7, 7.8, 7.10, 7.13, 7.14); *J. Am. chem. Soc.* (Copyright Am. Chem. Soc.) (Fig. 3.31); *Phys. Rev.* (Fig. 2.10, 2.22); *Rev. mod. Phys.* (Fig. 4.2, 4.9); Springer-Verlag from *Electrons in fluids* (Fig. 2.8, 2.14, 2.19, 7.1, 7.9, 7.12, 8.1, 8.8, 8.9); Université Catholique de Lille, from *Metal-ammonia solutions* (Fig. 3.25, 3.30).

CONTENTS

1

INTRODUCTION

1.1. Ammonia as a solvent

MANY solids combine with liquids to form homogeneous mixtures of only one phase; these mixtures are usually called solutions. Familiar examples include salt in water and sugar in tea. The solvent is generally taken to be the majority component and the solute or solutes the minority component. This monograph concerns a class of solutions in which the solute is an element: a metal. The most common solvent is liquid ammonia. Solutions of metals in ammonia differ from more common solutions in the extraordinarily wide range of values observed for many properties.

These materials have been studied for over a century, yet their nature is not resolved and controversy persists, even over such apparently simple properties as the density. This book does not attempt to resolve many of the discrepancies.

The archetypal solution for man is salt in water and the major thrust of most solution research has been the elucidation of the properties of similar solutions. When dealing with non-aqueous solutions, however, one must consider the differences between water and the solvent of interest. Several differences immediately appear when water and ammonia are compared. Water has the higher low-frequency relative permittivity (80 versus 22 at 20 °C) and the lower polarizability ($1 \cdot 45 \, \text{Å}^3$ versus $2 \cdot 26 \, \text{Å}^3$). Hydrogen bonding is more important in solvent–solvent interactions in H_2O than NH_3. Finally, the ammonia molecule is more stable against dissociation in the liquid than the water. In water the equilibrium concentrations of H^+ and OH^- ions are near $0 \cdot 1$ p.p.m. as compared to H^+ and NH_2^- concentrations of 10^{-10} p.p.m. in ammonia. On the other hand, the H_2O molecule is more stable in the vapour phase. There are, of course, major differences between metal–ammonia solutions and salt water due to the fact that the solutes discussed here are elements. The anion is thus an electron instead of a halide ion, charge transfer can occur without transfer of matter, and quantum effects can be important.

Perhaps most important to the understanding of these solutions, though certainly *not* well known, is the nature of the short-range electron–molecule interaction. Cohen, Jortner, and their co-workers have characterized the repulsive pseudopotential in liquids such as He, Ar etc. by the energy of an electron in the conduction band relative to the vacuum state. They find that for helium the band lies higher than the vacuum and for argon the band lies lower. Thus in the former case the electron can lower its energy by repelling the solvent and 'blowing itself a bubble' in the liquid, whereas in the latter the electron may move directly into the solvent. Both H_2O and NH_3 are isoelectronic with Ne, for which the band and vacuum states are very close. In the absence of a detailed calculation we can only guess that these polar solvents will have a weak repulsive interaction with electrons somewhat like that of Ne. Minor differences in energy may well produce significant changes in behaviour, as shall be seen when solvated electron spectra are discussed.

In the present chapter we briefly survey the early history of metal solutions in various solvents, then outline precisely the materials which are known to form metal solutions, and finally describe the preparation and preservation of metal solutions.

Recent application of M–NH_3 solutions has included the production of superconductors by the intercalation of e.g. Na into transition metal dichalcogenides (Acrivos, Liang, Wilson, and Yoffe 1971; Somoano and Rembaum 1971). They have been used in organic reactions for many years (Smith 1963) and in conventional inorganic chemistry (Watt 1957). These and other, more esoteric (Jäger and Lochte-Holtgreven 1967), uses of these materials are beyond the scope of this book.

1.2. Early history

The earliest known experiments in which a metal was observed to dissolve in ammonia were those reported by Weyl (1864). He studied sodium and potassium in ammonia and concluded that 'ammoniums' (NH_4) were being formed. (See also Joannis 1889.) Only seven years later Seely (1871) was able to recover Na dissolved in NH_3 by evaporating the solvent. He concluded that the metal did not react with the solvent and that the solution was similar to salt in water. No work was reported between 1873 and 1889. NH_3 became commercially available in 1898 and research was accelerated.

Further confirmation of these observations came from the 'Kansas group' of H. P. Cady, E. C. Franklin, and C. A. Kraus (Taft 1933). A series of experiments, beginning in 1895, was continued into the 1930s and provided the world with much of the available information on metal–ammonia solutions. Kraus was author or co-author of over fifty major papers on the solutions, and his students, or their students, are still major contributors. Phase diagrams, vapour pressures, and conductivities were among the properties studied by Kraus and his co-workers (Kraus 1931). His aim was to study the expected gradual change from electrolytic to metallic properties. He was able to conclude in 1908 (Kraus 1908) that the valence electron from the metal was associated with the solvent in dilute solutions yet free in concentrated solutions. He proposed that the electron in dilute solutions was surrounded by 'an envelope of solvent molecules'—a construct still in use (see Chapter 3). He realized at a very early date that the 'metallic electron' was involved and that the mobilities were very high (Kraus 1921), even by metallic standards. These deductions are remarkable in light of the poor understanding of metals at the time he worked. (The Drude model was only proposed in 1900.) Kruger (1938) was a proponent of the view that colloids existed rather than solutions. As we shall see this view does not fit the observations. Herzfeld (1927) pointed out that a rapid metal–nonmetal (M–NM) transition might exist (Cohen and Thompson 1968; Thompson 1968; Mott 1974) and that these solutions would be a good system in which to search for such a transition, as Kraus had foreseen. Mott's classic paper (1961) on the M–NM (or Mott) transition again drew attention to the metal–ammonia solutions as a system in which such a transition might exist.

1.3. Other sources of information

An extensive (perhaps exhaustive) bibliography of work on metal–ammonia solutions has been prepared by Professor G.

Lepoutre.† Further references to research can be found in the reports of conferences devoted to the subject.‡ Many important papers have been collected by Jolly (1972) and there have been several reviews (Kraus 1931; Jolly 1959; Symons 1959; Das 1962; Dye 1967; Cohen and Thompson 1968). We turn now to a survey of the materials involved, their preparation, and stability.

1.4. Accessible concentrations and temperatures

Though ammonia (NH_3) is the primary solvent used, other amines have also been reported to dissolve metals to some extent. Henceforth, except in the final chapter, we shall limit ourselves to a discussion of solutions of metals in liquid ammonia. In the last chapter (chapter 8) we shall take up solutions of metals in solvents such as methylamine, some ethers, etc. The emphasis will be on solutions of alkali metals as more information is available on them.

Liquid ammonia is known to dissolve the following metals: Li, Na, K, Rb, Cs, Ca, Sr, Ba, Yb, and Eu (Jolly 1959). One may also prepare solutions containing radical cations such as tetra-alkylammonium (Quinn and Lagowski 1968) by electrolysis. Finally, one may introduce electrons from an accelerator and study the anion without cationic effects (Dye 1968; Hart 1969). The alkali metals have been more extensively studied than the others and will be the major subject of all chapters except Chapter 6.

The present author has speculated (Thompson 1970) that the alkali metals and liquid ammonia will be miscible in all proportions at temperatures near the melting point of the metal. This speculation is based on the reported complete miscibility of Cs and NH_3 (Schroeder, Thompson, and Oertel 1969) and upon Kraus's observation that NH_3 is soluble in an alloy of Na with K (NaK) (Kraus 1907). Most experiments, however, are carried out

---

† G. Lepoutre, Laboratoire de Chimie-Physique, Faculté Catholique de Lille, Lille, Nord, France.

‡ *J. phys. Chem.* **57,** 547 *et seq.* (1953). Colloque Weyl I at Lille, France, 1963; published as *Metal–ammonia solutions,* ed. G. Lepoutre and M. J. Sienko, W. A. Benjamin, New York, 1964. 150th Meeting Am. chem. Soc., Atlantic City, New Jersey, 15–16 September 1965; published as *Solvated Electron, Adv. Chem Ser.* 1965, **50.** Colloque Weyl II at Ithaca, New York, June 1969, published as *Metal–ammonia solutions,* ed. J. J. Lagowski and M. J. Sienko, Butterworths, London, 1970. *Ber. Bunsenges. physik. Chem.* 1971, No. 7, **75.** Colloque Weyl III at Hanita, Israel, June 1972; published as *Electrons in Fluids,* ed. J. Jortner and N. R. Kestner, Springer-Verlag, Heidelberg, 1973. Colloque Weyl IV at East Lansing, Michigan, June 1975; published as *J. Phys. Chem.* **79,** No. 26 (1975).

TABLE 1.1

Solubilities (in m.p.m.[†]) at the n.b.p. (239·8 K) of ammonia

Metal	Li	Na	K	Rb	Cs	Ca	Sr	Ba	Eu	Yb
Solubility	20	16	15	—	65	16	16	16	—	16
Reference	(Jolly 1959; Lo 1966; Bridges, Ingle, and Bowen 1970)	(Jolly 1959)			(Schroeder *et al.* 1969)					

† For this paper, the concentration will be expressed either in mole fractions x, where $x = $ (moles metal)/(moles metal + moles solvent), or as mole percent metal (m.p.m.) = 100 x. In the case of divalent metals, the number of moles of metal may be multiplied by 2 to reflect the electron concentration. Such concentrations will be denoted 'electron concentration'. Other units include molarity and molality (Eggers, Gregory, Halsey, and Rabinovitch 1964).

near the normal boiling point (n.b.p.) of ammonia and solubilities for several metals at that temperature are given in Table 1.1. The addition of metal to liquid depressed the freezing point from its usual value of 195·4 K to temperatures as low as 89 K (Mammano and Coulter 1967). Table 1.2 collects the known eutectics.

TABLE 1.2

Eutectics of metal–ammonia solutions (Thompson 1967)

Metal	Li	Na	K	Rb	Cs	Ca	Sr	Ba	Eu	Yb
Eutectic concentration (m.p.m.)	20	17	15	—	15·7	13·3	13·3	13·5	—	14·3
Eutectic temperature (K)	88·6	163	116	—	98	185	185	186	—	183
Reference	(Mammano and Coulter 1969)				(Lelieur and Rigny 1973*b*)	(Schroeder *et al.* 1969)				

Experiments may thus be carried out over a wide range of composition and over a wide range of temperature, extending, perhaps, above the critical point (406 K) of ammonia (Naiditch 1964; Schindewolf 1968; Hart 1970; McNutt, Kinnison, and Ray 1974). The available range for lithium in ammonia is summarized in Fig. 1.1. Region I is the miscibility gap—solutions cannot be made at the temperatures and concentrations within that region

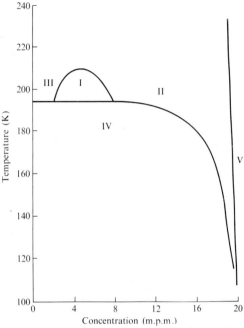

FIG. 1.1. Phase diagram for Li–NH₃ solutions. *Note* here and throughout the text m.p.m. = moles per cent metal (see p. 5).

(see Chapter 5); region II is a homogeneous, ·metallic liquid (Chapter 2); region III a homogeneous, nonmetallic liquid (Chapter 3); in region IV there is excess solid ammonia; and in region V there is excess solid metal or metal amine (Chapter 7). The phase diagrams are all quite similar, as may be seen in Figs 1.2 and 1.3 though Cs–NH₃ solutions lack region I (Schroeder *et al.* 1969).

1.5. Preparation of solutions

As noted earlier one is able to dissolve a variety of metals in liquid ammonia. In this section and in the next the preparation of the solutions and the measures required to keep the solutions for times sufficient to perform an experiment will be described.

Problems are derived from the combination of low temperatures required to maintain the NH₃ in the liquid state (though gaseous NH₃ will cause Li to deliquesce) and the inert atmospheres and extreme purities required to keep the solutions from decomposing. We will first describe the preparation of pure NH₃

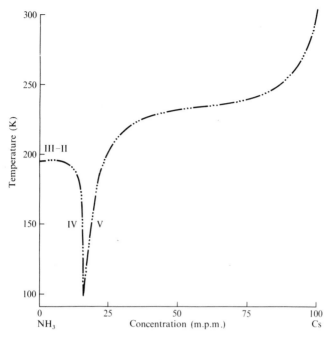

FIG. 1.2. Phase diagram for Cs–NH₃ solutions.

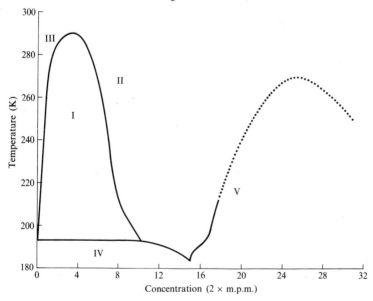

FIG. 1.3. Phase diagram for Ca–NH₃ solutions. The dotted portion is speculative (see Chapter 7).

and then the combination of the NH_3 with the metal of interest in the ratio desired.

Commercial non-aqueous ammonia contains 50 p.p.m. of H_2O (azeotrope) and is customarily stored at room temperature and pressures of 150 p.s.i. gauge. One often uses it at 240 K or below, and the water content must be reduced by several orders of magnitude. Simple distillation is not enough. The most effective procedure has been described by Dewald (Dewald and Roberts 1968). Pure dry NH_3 is first condensed in a trap containing sodium metal and stored for a day at 240 K. Water present in the ammonia is thus allowed to react with the metal. The ammonia is then distilled into the sample cell and back. The solution is next frozen with liquid nitrogen and pumped to remove dissolved hydrogen and the sample cell flamed under vacuum. This cycle is repeated several times. The first product has a conductivity below $10^{-7} \, \Omega^{-1} \, cm^{-1}$. For lower conductivities this kind of procedure must be repeated perhaps 30 times (Clutter and Swift 1968) until concentrations of impurities such as Na^+ or OH^- are below 10^{-6} molar. Conductivities below $10^{-10} \, \Omega^{-1} \, cm^{-1}$ have been reported. Na may be leached from the glass by some solvents (Hurley, Tuttle, and Golden 1970).

Most of the alkali metals can be cleaned by multiple distillation. Lithium must be mechanically cleaned in an inert atmosphere if large amounts are required, otherwise Dye's procedure of evaporating the metal onto a cold substrate may be used. The alkali earth metals also are cleaned only mechanically and rapid decomposition is the result. Plummer and Senozan (1971) report Sr to be easiest of the divalent metals to handle.

It is also not enough to clean only the NH_3 and the metal. Contaminants will also be found on or in the surface of the glass or quartz container (Feldman, Dewald, and Dye 1965; Warshawsky 1963; Naiditch and Wreede 1965; Burow and Lagowski 1965; Jackman and Keenan 1968).

Na atoms and H_2O molecules as well as radicals such as OH^- often enter the solutions to act as impurities or reactants. Warshawsky (1963) believes that unspecified 'sites' on the surface of the containing vessel catalyze the reaction:

$$NH_3 + e^- \rightarrow NH_2^- + \tfrac{1}{2}H_2, \qquad (1.1)$$

which removes solvated electrons from the solution (Kirschke and Jolly 1967; Jolly 1970). The amide (NH_2^- + metal cation) may form a film covering the surface of the containing vessel and reducing the contact between solute and reasiting 'sites' and thus lowering the reaction rate. Appreciable reaction can occur within a few minutes after the solution is made. From the nature of Warshawsky's observations one must expect that the initial reaction of metal with surface contaminants occurs to some extent in all experiments. If the solution's concentration is determined by a total metal determination and by total solvent determination (e.g. when metal and solvent are weighed prior to mixing), the composition so obtained must be in error. Neither the simple constancy of a parameter, such as conductivity, with time nor an extrapolation of a time-varying parameter to the time of mixing can show the absence of decomposition or eliminate the effects of decomposition. The error is fixed by the surface left by the cleaning procedures used and is therefore likely to be the same for measurements from a given laboratory or apparatus. Furthermore, the error in the metal concentration is not a given fraction of the metal used, rather it is a given amount of metal since the extent of the reaction depends on the surface available. This point makes such errors vastly more important in dilute solutions.

Feldman *et al.* (1965) have proposed the following cleaning procedure for glassware. First, the items are rinsed briefly with a hydrofluoric acid–detergent cleaning solution (5% hydrofluoric acid, 33% concentrated reagent grade nitric acid, 2% acid-soluble detergent, and 60% distilled water), then ten times or more with distilled water. Next, they are filled with fresh *aqua regia* and heated to boiling, followed by ten more rinses with distilled water, then by several rinses and/or by soaking in double-distilled conductance water. After each use and a preliminary cleaning, they are heated in an annealing oven at 550 °C before being cleaned as described above. The multiple-rinse techniques have been developed specifically to replace such species as H_2O with NH_3 and to react with as much of any other species as possible. Burow and Lagowski (1965) report that using a dilute metal–ammonia solution for the washing agent is particularly effective.

Few containers other than glass or quartz are available; even quartz induces more decomposition than borosilicate glass. Schindewolf, Böddeker, and Vogelsgesang (1966) and also Varlashkin and Stewart (1966) have used steel and stainless steel containers, respectively. Schindewolf found some improvement when the steel was covered with gold plate. The only plastic known to be a good container is polyethylene (Mammano and Coulter 1969; Brady and Varimbi 1964). Though Teflon is commonly considered inert, these solutions are used commercially to etch Teflon! Mylar is also rapidly attacked. No cleaning procedures equivalent to that of Feldman *et al.* have been developed for metallic or plastic systems. Aluminium is said to tend to shift reaction (1.1) to the left (P. Varlashkin, private communication).

Cleansing procedures must be more stringent for the heavier elements e.g. Cs. Reactions such as (1.1) which remove electrons are naturally more damaging when the metal content is low. If electrodes are present, care is also necessary for their cleansing (Dewald and Roberts 1968; Nasby and Thompson 1968). Several other authors have reported the influence of clean apparatus and materials on the results of experiments (Dewald and Lepoutre 1954, 1956; Jolly 1965; Gunn 1967). In all cases lower temperatures inhibit decomposition and permit experiment times to be extended.

Most of the references above will yield information on 'typical' procedures and systems; recent work of Rosenthal and Maxfield (1973) describes a complete system in detail.

The careful reader will have noted that most of the references to cleaning procedures are very recent, for the influence of decomposition has been appreciated only over the past few years. As a consequence much of the older data, particularly in the most dilute range (Kraus 1921; Dewald and Roberts 1968) is suspect. Though few qualitative changes are expected as experiments are repeated there have already been quantitative changes in the results which produce formidable problems for the would-be theorist. The complexity of the solutions seems to increase with the precision of the experiment. The chapters following present the data together with a critical comparison of data and theory. The problem will be seen still to be open.

REFERENCES

Acrivos, J. V., Laing, W. Y., Wilson, J. A., and Yoffe, A. D. (1971). *J. phys. Chem.* **3,** 118.

BRADY, G. W. and VARIMBI, J. (1964). *J. chem. Phys.* **40**, 2615.
BUROW, D. F. and LAGOWSKI, J. J. (1965). *Adv. Chem. Ser.* **50**, 125.
CLUTTER, D. R. and SWIFT, T. J. (1968). *J. Am. chem. Soc.* **90**, 601.
COHEN, M. H. and THOMPSON, J. C. (1968). *Adv. Phys.* **17**, 857.
DAS, T. P. (1962). *Adv. chem. Phys.* **4**, 303.
DEWALD, J. F. and LEPOUTRE, G. (1954). *J. Am. chem. Soc.* **76**, 3369.
—— —— (1956). *J. Am. chem. Soc.* **78**, 2956.
DEWALD, R. R. and ROBERTS, J. H. (1968). *J. phys. Chem.* **72**, 4224.
DYE, J. L. (1967). *Scient. Am.* **216**, 77.
—— (1968). *Acc. chem. Res.* **1**, 306.
EGGERS, D. F., GREGORY, N. W., HALSEY, G. D., and RABINOVITCH, B. S. (1964). *Physical chemistry.* Wiley, New York.
FELDMAN, L. H., DEWALD, R. R. and DYE, J. L. (1965). *Adv. Chem. Ser.* **50**, 163.
GUNN, S. R. (1967). *J. chem. Phys.* **47**, 1174.
HART, E. J. (1969). *Acc. chem. Res.* **2**, 161.
—— (1970). In *Metal–ammonia solutions* (ed. J. J. Lagowski and M. J. Sienko) p. 413. Butterworths, London.
HERZFELD, K. M. (1927). *Phys. Rev.* **29**, 701.
HURLEY, I., TUTTLE, T. R., and GOLDEN, S. (1970). In *Metal–ammonia solutions* (ed. J. J. Lagowski and M. J. Sienko), p. 449. Butterworths, London.
JACKMAN, D. C. and KEENAN, C. W. (1968). *J. inorg. nucl. Chem.* **30**, 2047.
JÄGER, H. and LOCHTE-HOLTGREVEN, W. (1967). *Z. Phys.* **198**, 351.
JOANNIS, A. (1889). *C.r. hedral. Seanc. Acad. Sci. Paris.* **109**, 900; 965.
JOLLY, W. L. (1959). *Prog. inorg. Chem.* **1**, 235.
—— (1965). *Adv. Chem. Ser.* **50**, 27.
—— (1970). In *Metal–ammonia solutions* (ed. J. J. Lagowski and M. J. Sienko), p. 167. Butterworths, London.
—— (1972). *Metal ammonia solutions.* Dowden, Hutchinson Ross, Stroudsburg, Pennsylvania.
KIRSCHKE, E. J. and JOLLY, W. L. (1967). *Inorg. Chem.* **6**, 885.
KRAUS, C. A. (1907). *J. Am. chem. Soc.* **29**, 1557.
—— (1908). *J. Am. chem. Soc.* **30**, 653.
—— (1921). *J. Am. chem. Soc.* **43**, 741.
—— (1931). *J. Franklin Inst.* **212**, 537.
KRUGER, F. (1938). *Annln. Phys.* **33**, 265.
LELIEUR, J. P. and RIGNY, P. (1973). *J. chem. Phys.* **59**, 1148.
LO, R. E. (1966). *Z. anorg. Allg. chem.* **344**, 230.
MAMMANO, N. and COULTER, L. V. (1967). *J. chem. Phys.* **47**, 1564.
—— —— (1969). *J. chem. Phys.* **50**, 393.
McNUTT, J. D., KINNISON, W. W., and RAY, A. D. (1974). *J. chem. Phys.* **60**, 4370.
MOTT, N. F. (1961). *Phil. Mag.* **6**, 287.
—— (1974). *Phil. Mag.* **29**, 613.
NAIDITCH, S. (1964). *Metal–ammonia solutions*, p. 113. (ed. G. Lepoutre and M. J. Sienko) W. A. Benjamin, New York.
—— and WREEDE, J. (1965). Final Technical Report, Contract NONR3437 (00). Unified Science Associates, Pasadena, California. (Unpublished).
NASBY, R. D. and THOMPSON, J. C. (1968). *J. chem. Phys.* **49**, 969.
PLUMMER, G. and SENOZAN, N. M. (1970). *J. chem. Phys.* **55**, 4062.
QUINN, R. K. and LAGOWSKI, J. J. (1968). *J. phys. Chem.* **72**, 1374.
ROSENTHAL, M. D. and MAXFIELD, B. W. (1973). *J. Solid State Chem.* **7**, 109.
SCHINDEWOLF, U. (1968). *Angew. Chem.* **7**, 190.

SCHINDEWOLF, U., BÖDDEKER, K. W., and VOGELSGESANG, R. (1966). *Ber. (Dtsch) Bunsenges. phys. Chem.* **70,** 1161.

SCHROEDER, R. L., THOMPSON, J. C., and OERTEL, P. L. (1969). *Phys. Rev.* **178,** 298.

SEELY, C. A. (1871). *Chem. News* **23,** 169.

SMITH, H. (1963). *Organic reactions in liquid ammonia.* Interscience, New York.

SOMOANO, R. B. and REMBAUM, A. (1971). *Phys. Rev. Letts.* **27,** 402.

SYMONS, M. C. R. (1959). *J. chem. Phys.* **30,** 1628–9.

TAFT, R. (1933). *J. chem. Educ.* **10,** 196.

THOMPSON, J. C. (1967). In *Chemistry of non-aqueous solvents* (ed. J. J. Lagowski) p. 265. Academic Press, New York.

—— (1968). *Rev. mod. Phys.* **40,** 704.

—— (1970. In *Metal–ammonia solutions* (ed. J. J. Lagowski and M. J. Sienko) p. 292. Butterworths, London.

VARLASHKIN, P. G. and STEWART, A. T. (1966). *Phys. Rev.* **148,** 459.

WARSHAWSKY, I. (1963). *J. inorg. nucl. Chem.* **25,** 601.

WATT, G. W. (1957). *J. chem. Educ.* **34,** 538.

WEYL, W. (1864). *Poggendorffs Annln.* **121,** 601.

2

METAL–AMMONIA SOLUTIONS AS LIQUID METALS

2.1. Introduction

A SATURATED solution of lithium in liquid ammonia has an electrical conductivity of $15\,000\,\Omega^{-1}\,\text{cm}^{-1}$, which is higher than that of liquid mercury at room temperature (Morgan, Schroeder and Thompson 1965). Other metals dissolved in liquid ammonia have conductivities almost as high (Kraus 1907, 1921a, 1921b, 1931; Kraus and Lucasse 1921, 1922, 1923; Schroeder, Thompson, and Oertel 1969) and the addition of ammonia to such solutions does not immediately reduce the conductivity beyond the range commonly associated with metallic behaviour (Allgaier 1969). For purposes of discussion and for reasons best given later, solutions containing more than 8 mole per cent of an alkali metal will be considered as liquid metals. As will be seen, one must not infer from this that these solutions behave as do the pure alkali metals when liquefied rather than dissolved. In this chapter the experimental and theoretical basis for the assertion that these solutions do indeed behave as liquid metals will be examined. The subject of the transition to the nonmetallic state is referred to in Chapter 4. First, the properties of metallic solutions will be presented. Next, examples of the behaviour of liquids commonly regarded as metallic (alloys as well as single component systems) will be given together with a brief description of the theory of such conductors. The behaviour of the solutions will then be compared to that of the common liquid metals to give an indication of the problem to be solved. A model based on the Ziman theory of liquid metals will be given. A critique of models suggested previously follows. Finally, a statement of the presently most attractive model will be given. As in all discussions of metal–ammonia solutions, one can be almost overwhelmed by the amount of available data. The properties of the solutions known in this concentration range include those shown in Table 2.1.

TABLE 2.1

Properties of M–NH₃ solutions studied in the concentration range 8 m.p.m.-saturation

Transport coefficients	Magnetic effects	Mechanical properties	Other
σ(conductivity), $d\sigma/dT$, $d\sigma/dP$, $d\sigma/dB$	$k(H)$, $k(N)$, $k(M)$, and temperature coefficient (Knight shift)	ρ(density) β_s, $d\beta_s/dT$(adiabatic compressibility)	$N(k)$ (momentum distribution)e^+
S(thermopower), dS/dT	χ(susceptibility), $d\chi/dT$, χ_{Pauli}	β_T, $d\beta_T/dT$	τ(lifetime)e^+
κ(thermal conductivity), $d\kappa/dT$	e.s.r. line width and asymmetry	C_p	photoelectric threshold
R_H(Hall coefficient), dR_H/dT	T_1(spin–lattice relaxation time)(e), $T_1(H)$	η(viscosity), $d\eta/dT$	X-ray
$\varepsilon = \varepsilon_1 + i\varepsilon_2$(optical)		Σ(surface tension) $d[\Sigma]/dT$	

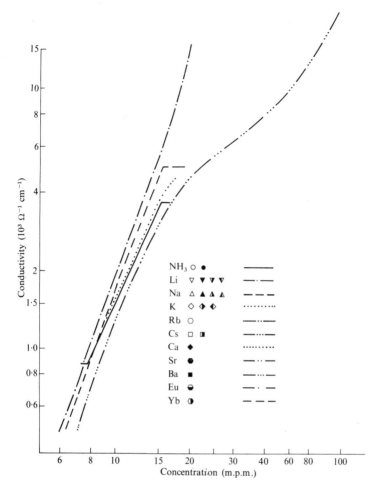

FIG. 2.1. Electrical conductivities of solutions of several metals in liquid ammonia at 240 K. (Schroeder, Thompson, and Oertel 1969). In this and subsequent figures data pertaining to specific solutes or to NH₃ are represented by a consistent set of symbols (whenever possible) shown on the figure. The open and filled symbols for Li, Na, or K are used whenever it is desirable to separate the work of different authors; for NH₃ the two symbols may denote different authors or different solutes. For this figure only, the Ca–NH₃ conductivity is denoted by a solid line.

There is obviously a wealth of data on which to build a model of the solutions. The list of properties given is misleading, however, as the solute, the temperature, or the care exercised varies significantly from experimenter to experimenter.

In the interpretation of the data in this range, as in solutions containing less metal, one is guided by the observation that the differences in most properties produced by substituting one alkali metal for another are not exceedingly important, and that these differences are much smaller than between the corresponding pure alkali metals. The most widely determined property of the solutions is the electrical conductivity (Kraus 1931; Schroeder *et al.* 1969). Fig. 2.1 shows the electrical conductivity of several of the nine metals which have been studied in liquid ammonia. The graph shows two distinct classes: (1) the alkali metals; (2) the alkaline earth and rare earth metals. Within each class the deviations are small until one is close to saturation. Between the two groups the differences are not large though there appears to be a difference in the slope of the curves. Another property which has been fairly widely studied is the adiabatic compressibility (Maybury and Coulter 1951; Bowen, Thompson and Millett 1968; Bowen 1969; Bridges, Ingle, and Bowen 1970; Thompson

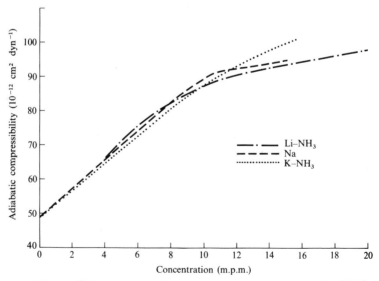

FIG. 2.2. The adiabatic compressibilities of several M–NH₃ solutions at 233 K (Bowen 1970).

and Oré-Oré 1971). Fig. 2.2 shows some data for the compressibility and again shows remarkable little change from one solute to the next. Positron annihilation data (angular correlation of annihilation gamma rays) are independent of solute (Varlashkin and Stewart 1966; Varlashkin 1968; McCormack and Millett 1966; Arias-Limonta and Varlashkin 1970). Further examples of the similarity of the solutions will occur as other properties are discussed both in the present metallic concentration range and in the more dilute concentration range to be discussed in Chapter 3.

2.2. Properties of concentrated M–NH₃ solutions

In the following paragraphs, we will systematically examine the available data in the concentration range above 8 m.p.m. though we shall defer the discussion of alkaline earth and rare earth metal solutions until the end. To avoid undue presentation of graphical material, only data judged typical of each of the various properties will be shown.

2.2.1. *Electrical conductivity*

The electrical conductivity has been measured in solutions of lithium, sodium, potassium, rubidium, and caesium in liquid ammonia over the concentration range from 8 m.p.m. to saturation and at one or more temperatures. Some of these data are a half century old, others have been obtained since 1970. The techniques used vary from conventional four-probe d.c. measurements (Morgan *et al.* 1965; Kraus 1907, 1921*a, b*) to more elaborate techniques dependent upon the decay of eddy currents (Morgan *et al.* 1965; Schroeder *et al.* 1969, Castel *et al.* 1971). It is difficult to compare the quality of the data because many authors use conductivity measurements to determine the concentration. Kraus's data (the oldest) have generally been verified in this range when verification has been attempted. Small differences between electrodeless and other measurements have been attributed to decomposition at the electrodes (Morgan *et al.* 1965). It seems unlikely that major errors exist though there is need for verification of some of the data (particularly that on alkaline earth metal solutions). In most cases data is also available on the temperature coefficient of conductivity. In addition there has been one examination of the influence of hydrostatic pressure on the conductivity of sodium–ammonia solutions (Schindewolf, Boddeker, and Vogelgesang 1966) as well as a search for a change in conductivity with magnetic field in lithium–ammonia solutions (Kyser and Thompson 1964).

The data on the electrical conductivity of several lithium–ammonia solutions is presented in Fig. 2.3 in somewhat greater detail than in Fig. 2.1. The following observations should be made. First, the electrical conductivity increases with concentration approximately as the cube of the metal content (Schroeder *et al.* 1969). As will be seen, the density of metal ions is proportional to the metal content so there is also an approximately

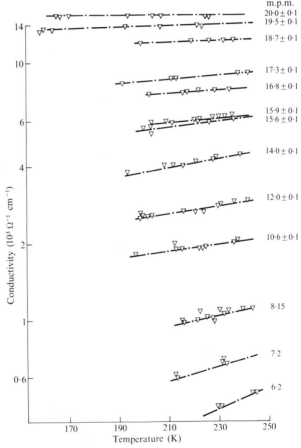

FIG. 2.3. Electrical conductivities of several Li–NH₃ solutions as a function of temperature (Schroeder, Thompson, and Oertel 1969). The concentrations are listed at the right in m.p.m.

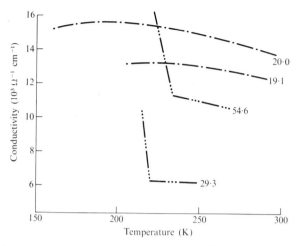

FIG. 2.4. Electrical conductivities of some M–NH₃ solutions as a function of temperature (Schroeder *et al.* 1969). Note that there is a maximum for the Li–NH₃ data. The abrupt change of slope indicates a phase boundary in the Cs–NH₃ data and such data have been used in the construction of Fig. 1.2.

cubical dependence of electrical conductivity upon valence electron density in the alkali metal solutions since each metal ion implies the presence of a dissociated electron. Secondly, the electrical conductivity is an increasing function of temperature except for the most concentrated lithium or caesium solutions (Castel *et al.* 1971) wherein the conductivity has an extremum and begins to decrease slightly with temperature as shown in Fig. 2.4. In some Li–NH₃ solutions the linear increase of σ with T has been observed over a 200 degree range; in most cases the range studied is only 30 degrees. Caesium solutions, which do not saturate (that is, caesium and ammonia are miscible in all proportions) show above 30 m.p.m. a conductivity which is a decreasing function of temperature; in the main, remarks will be restricted to the concentration range common to nearly all the metals: 8–20 m.p.m. Therein, a general feature of these solutions is a conductivity which increases as the temperature increases. The increase is very closely linear with temperature (Nasby and Thompson 1970). Thirdly, the application of pressure reduces the conductivity of the solutions (Schindewolf *et al.* 1966). Again the effect is approximately linear in the pressure and (as also observed in the temperature coefficient) the effect of pressure is

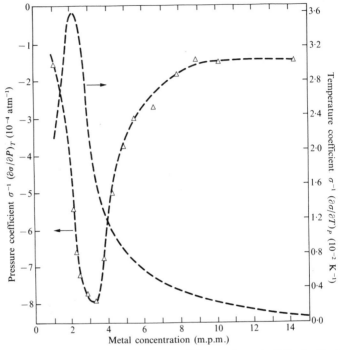

FIG. 2.5. Temperature and pressure coefficients of conductivity in Na–NH₃ solutions near 233 K. Note that $\partial\sigma/\partial T$ is positive while $\partial\sigma/\partial P$ is negative, and very large near 4 m.p.m. (Schindewolf *et al.* 1966).

lessened as the concentration approaches saturation. Fig. 2.5 collects the temperature and pressure coefficients of conductivity in terms of the appropriate logarithmic derivative of the conductivity and shows most clearly this decreasing effect with increasing concentration. Fourthly there has been no observed effect of magnetic field upon conductivity in those metal–ammonia solutions which have been examined (Kyser and Thompson 1964).

In summary then, the electrical conductivity is an increasing function of concentration, an increasing function of temperature, and a decreasing function of pressure. It is independent of applied magnetic field. Solute effects are small.

2.2.2. *Hall effect*

The Hall effect has been studied in solutions of lithium, sodium, potassium, and calcium in liquid ammonia. (Kyser and

Thompson 1965; Nasby and Thompson 1970; Naiditch, private communication; Vanderhoff and Thompson 1971). In each case the experimental technique involved the use of two a.c. frequencies. That is, an a.c. current and an a.c. magnetic field were used, with the Hall voltage appearing as the sum or difference frequency of the two applied signals. This technique was adopted to reduce the motion of the conducting liquid in the applied magnetic field (Russell and Wahlig 1950; McKinzie and Tannhauser 1969). Because of the various problems of pick-up and of non-linearities which have been discussed by Suchannek (1966) and by Nasby and Thompson (1968), these measurements are not at all simple and the data are not as accurate as one would like. Nevertheless, one may draw the following conclusion: the carrier density in metal–ammonia solutions containing more than 8 mole per cent of an alkali metal is equal to the density of metal valence electrons and is independent of temperature, except as the density of the solution varies with temperature. Fig. 2.6 shows the Hall coefficient of lithium–ammonia solutions as measured by Nasby and Thompson (1970). In fact, it has been primarily the Hall data which have forced the conclusion that the concentrated solutions are liquid metals, with no carrier activation.

The product of the conductivity σ and the Hall coefficient R_H is the Hall mobility μ_H, shown in Fig. 2.7. There remains a strong concentration and temperature dependence even of μ_H.

2.2.3. Thermoelectric power, S

The thermoelectric power S in the concentration range presently examined, has been measured by Dewald and Lepoutre (1954, 1956), and by Damay, Depoorter, Chieux, and Lepoutre (1970). As Dewald and Lepoutre's experiments have been cited as examples of the care which should be taken in preparation of solutions, one cannot question these data on the basis of decomposition as we can in some other experiments. The electrode problems found by Nasby and Thompson (1968) and by Schettler and Patterson (1970) may have influenced the data in concentration ranges lower than those presently considered. The raw data must be corrected for the thermopower of the conducting leads to the measuring instruments (MacDonald 1962) and when this is done one finds that there is a change in the sign of the thermoelectric

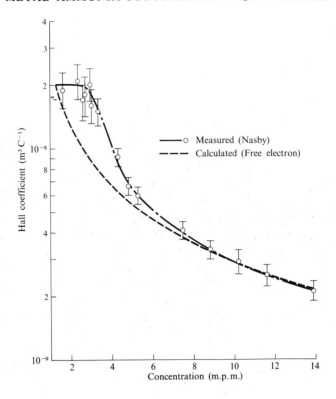

FIG. 2.6. Measured Hall coefficient for Li–NH₃ solutions compared to the value expected from free electron theory (dashed line) (Nasby and Thompson 1970).

power, of sodium solutions at least, near a concentration of 13 m.p.m. This change of sign is apparently a real phenomena and not a consequence of stray e.m.f.s in the circuit, as it persists when the temperature of the experiment is changed. There is no change of sign in K–NH₃ solutions.

The magnitude, though not the sign, of the thermoelectric power is appropriate for a metallic conductor, being small and approximately equal to that to be expected on the basis of free electron theory as is shown in Fig. 2.8. In that figure the calculated thermoelectric power is based on the electron density determined by Nasby and Thompson (1970) using the Hall effect, and the thermoelectric power data is due to Dewald and Lepoutre (1954, 1956). Note, however, that the Hall data were

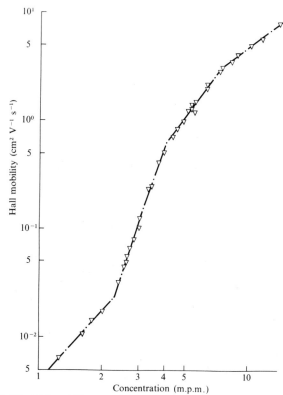

FIG. 2.7. The Hall mobility $\mu_H \equiv R_H \sigma$ for Li–NH$_3$ solutions at 233 K (Nasby and Thompson 1968).

taken on lithium and the thermoelectric power data were taken on potassium. Damay, Depoorter, Chieux, and Lepoutre (1970) have recently reported that the thermoelectric power is a decreasing function of temperature.

2.2.4. Thermal conductivity

Varlashkin and Thompson (1963) measured thermal conductivity of solutions of lithium, sodium, and potassium in liquid ammonia. Here again, we are faced with an exceedingly difficult experiment where problems of decomposition and of accounting for container and other losses can make the interpretation of the data extremely ambiguous. Varlashkin found a thermal conductivity in all cases which was slightly smaller in magnitude than that to be expected from the Wiedemann–Franz Law (Ziman

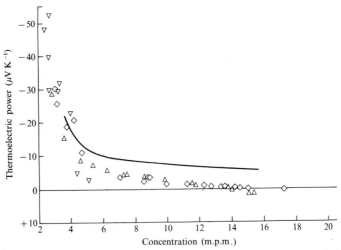

FIG. 2.8. The absolute thermoelectric power for several M–NH₃ solutions (Damay *et al.* 1970) at 240 K.

1960; Cohen and Thompson 1968) and an increase in thermal conductivity with temperature which was approximately that to be expected on the basis of Wiedemann–Franz Law. Fig. 2.9 shows the Lorenz number as a function of concentration for some of Varlashkin's data and we see that the measured Lorenz number is smaller than that to be expected on free electron

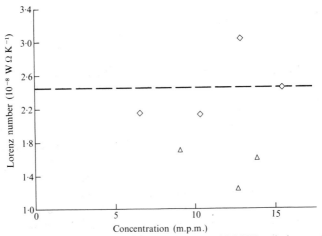

FIG. 2.9. The Lorentz number $L \equiv K/\sigma T$ for several M–NH₃ solutions at 235 K (Cohen and Thompson 1968).

theory. These data again are strongly in need of verification by independent experimenters. There is internal consistency between the electrical and thermal conductivities in that the variation with composition and temperature is approximately that given by the Wiedemann–Franz Law, showing the essentially electronic nature of both processes.

2.2.5. Optical data

Optical data have been reported by Beckman and Pitzer (1961), by Stanek and Lagowski (1969) and by Thompson and his co workers (Thompson and Cronenwett 1967; Somoano and Thompson 1970; Mueller and Thompson 1970; Vanderhoff, LeMaster, McKnight, Thompson, and Antoniewicz, 1971). The first-named authors measured the normal reflectance relative to that of mercury. Their data covered too narrow a wave length range to permit extraction of more interesting parameters such as the optical constants. The data did show qualitative similarities to simple metals. Cronenwett and Thompson used an ellipsometric technique to measure the real (ε_1) and imaginary (ε_2) parts of the dielectric constant of the solutions as a function of concentration at 210 K. Again the problems of decomposition arise, made more severe in the case of the optical experiment by the fact that the observations are dependent upon the condition of the surface of the solutions or, more properly in Cronenwett's experiment, of the interface between the solution and a quartz prism. Nevertheless, the course of the experiment with time led Cronenwett to believe that decomposition did not unduly influence the optical constants which he reported. The influence of whatever experimental errors produced the Mayer–El Naby (1963) resonance cannot be judged here, though no structure was observed in the optical constants. Nevertheless, the complications which may derive from non-normal incidence may well have been present. Cronenwett did not attempt an analysis of such factors. It is at first surprising that he found ε_2 to be concentration independent. The raw data show changes with concentration which largely cancel out in the analysis. Observations with somewhat different techniques would be of interest. Vanderhoff *et al.* (1971) have re-examined the Li–NH$_3$ solutions using a reflection technique (Mueller 1969; Mueller and Thompson 1970) and found improved agreement with Drude theory. They also extended the data out to 5 eV without observing interband transitions. They did find indication of an amide impurity in the spectra at 4·2 eV (Caruso, Takemoto, and Lagowski 1968). A step in the preparation of samples was also changed. Previous experimenters had prepared the solutions in the optical cell, a process which sometimes left a cloudy surface. Vanderhoff *et al.* (1971) made up the solution in a different cell, mixed it well, then poured it into the optical cell. This produced a visually brighter surface.

Some data from Vanderhoff *et al.* (1971) are shown in Fig. 2.10, where they are compared with results to be expected on the

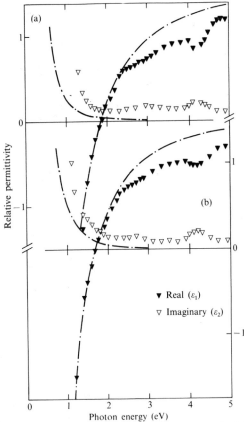

FIG. 2.10. Real and imaginary parts of the relative permittivity (dielectric constant) of Li–NH₃ solutions at 195 K (Vanderhoff *et al.* 1971). Curve (a) is for a 19·4 m.p.m. solution, while curve (b) is for a 15·2 m.p.m. solution. The dash–dot curves show the results expected from free electron theory.

basis of the Drude (1900) theory (Ehrenreich 1966; Pines 1963). The real part of the optical constant is typical of that seen in metallic systems, as is the imaginary part. There is an excess in the observed ε_2 at low energies over the Drude prediction. This shoulder was not observed in ellipsometric measurements (Somoano and Thompson 1970). The point at which the real part of the relative permittivity goes through zero is customarily taken to be the point of a collective resonance, the plasma resonance (Pines 1962) of the electrons in the material and there is qualitative agreement between the data and the Drude theory. If one computes the imaginary part of the reciprocal of the relative permittivity, $\mathrm{Im}(1/\varepsilon)$, one has a measure of the energy loss function (e.l.f.) and the peak of that function is related to the plasma frequency. The relations are given by:

$$\varepsilon = \varepsilon_1 - \mathrm{i}\varepsilon_2 = 1 + \frac{ne^2}{m}\left(-\omega^2 + \mathrm{i}\omega/\tau_{\mathrm{opt.}}\right)^{-1} \qquad (2.1)$$

$$\mathrm{e.l.f.} = \mathrm{Im}(1/\varepsilon) = \varepsilon^2/(\varepsilon_1^2 + \varepsilon_2^2) \qquad (2.2)$$

where n is the carrier density, ω the photon frequency, and $\tau_{\mathrm{opt.}}$ the optical relaxation. Values of the electron density n and electron relaxation time $\tau_{\mathrm{opt.}}$ obtained from the optical data are compared with those obtained by Nasby and Thompson (1970) from the Hall coefficient in Table 2.2. One sees there that certainly order-of-magnitude agreement exists between the quantities involved. Optical values of relaxation times significantly smaller than those obtained by d.c. measurements are typically found in other metallic systems (Abeles 1966; 1972). The effect of temperature upon the optical properties is small as would be expected from Drude theory and from d.c. values of the conductivity.

The above data involve the various transport coefficients, all of which are in some sense or another related to the dependence of the mean free path on concentration. Their behaviour is similar to that observed in simple liquid metals (Allgaier 1969; Cusack 1963; Faber 1972) except that $d\sigma/dx$ is large and $d\sigma/dT$ positive. Even these parameters are within the limits suggested by Allgaier (1969) for conduction by electron propagation rather than hopping. Allgaier (1969, 1970a), and Thompson and Allgaier (1970) (see also Thompson 1973) have tabulated and examined measurements of σ, R_{H}, and S for many electronically conducting

TABLE 2.2

Values of electron density derived from Hall[a] and optical data[b]

Concentration (m.p.m.)	n_H, from Hall data (cm^{-3})	n_0, from optical data (cm^{-3})	n_H/n_0	τ_H (10^{-15} s)	τ_0 (10^{-15} s)
4	0.59×10^{21}	0.65×10^{21}	0.92	0.32	—
5	0.964	0.89	1.08	0.56	—
8	1.69	1.39	1.22	1.9	~1.1
12	2.58	2.07	1.25	3.4	1.6
20	4.00	3.22	1.24	13.0	3.0

[a] Nasby and Thompson (1970).   [b] Thompson and Cronenwett (1967).

fluids. Allgaier describes three basic types. Class A fluids are good metals having σ in excess of $5000\ \Omega^{-1}\,cm^{-1}$ (perhaps lower if the density is low), are exemplified by Na, and may be described by nearly free electron (N.F.E.) theory (Ziman 1967). The mean free path L is long enough for the concept to be meaningful and the electron–ion interaction weak. Most conduction parameters have values and signs typical of solid metals. Class B liquids have σ between 100 and $5000\ \Omega^{-1}\,cm^{-1}$, and in most cases $d\sigma/dT$ is positive (as in semiconductors), though there is no other indication of thermal activation of carriers. In Class C liquids, non-metallic behaviour occurs. The boundary, $\sigma\sim 100\ \Omega^{-1}\,cm^{-1}$, between Classes B and C is an estimate of the lowest conductivity values for which it is still possible to describe charge transport by conventional extended electronic states. In Class C, this idea breaks down and models are based on hopping or tunnelling of electrons between localized states at the Fermi energy or alternatively are semiconductor models assuming activation to the mobility edge (Section 4.5.5). This subject will be pursued in Chapter 4. Most concentrated M–NH$_3$ solutions fall into Class B, only highly concentrated Li–NH$_3$ and Cs–NH$_3$ (Schroeder *et al.* 1969) solutions attain Class A. Allgaier (1970) also finds a 'universal' curve relating μ_H to σ. The M–NH$_3$ solution data do *not* fall on this curve. Next is a discussion of magnetic and other properties which are not so directly related to the mean free path.

2.2.6. *Magnetic susceptibility*

Measurements of the magnetic susceptibility, using a static method, have been performed in metal–ammonia solutions by Huster (1938), Freed and Sugarman (1943) and more recently by Suchannek, Naiditch, and Klejnot (1967), and Lelieur and Rigny (1973 a). These last experiments show a susceptibility which is weakly dependent on temperature. There is a temperature dependent susceptibility even in the most concentrated solutions. Some problems arise in the analysis of static susceptibility data. Liquid ammonia is diamagnetic; the addition of metal only makes the liquid less diamagnetic. The effect of the solute is separated from that of the solvent by the use of Wiedemann's rule (Myers 1952). This rule states that the susceptibility of a solution χ_s is

given by $\chi_s = (1-x)\chi_A + x\chi_B$ where χ_A and χ_B are the susceptibilities of solvent and solute respectively. The rule originates in
the presumed weak interaction of solvent and solute and has
been attacked in the case of small solute ions such as Li^+.
However the reliability of the static data is supported by the
closeness of static and spin susceptibilities in dilute solutions
(Hutchison and Pastor 1953). Lelieur and Rigny (1973a) have
used a resonance technique, due to Schumacher and Slichter
(1956), which compares electronic and nuclear susceptibilities,
and thereby confirmed their static measurements. There are
nevertheless disappointingly few data points (Graper and
Naiditch 1969). Suchannek et al. have a few data points near 10
and 12 m.p.m., but at room temperature, while Lelieur and Rigny
(1973a) report data points at only three concentrations in the 10
to 13 m.p.m. range along with the temperature coefficient. While
these data appear internally consistent, there is clear need for
further precise measurements. Atomic susceptibilities extracted
from a smooth curve drawn through the Lelieur and Rigny data
are shown in Fig. 2.11 and the temperature dependence is shown
in Fig. 2.12.

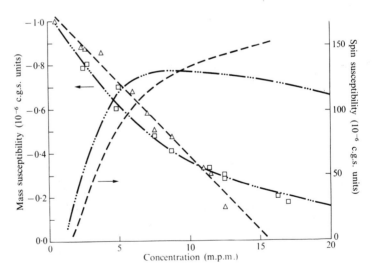

FIG. 2.11 Mass and spin susceptibilities for Na– and Cs–NH₃ solutions. (Lelieur
and Rigny 1973a, b). Data points are shown for the mass data while lines only are
given for spin data.

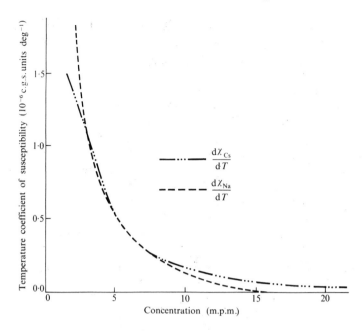

FIG. 2.12. Temperature derivative of spin susceptibility: taken from same source as Fig. 2.11.

The determination of the paramagnetic susceptibility in metallic solutions is not an easy problem. When the electrical conductivity is high enough, the skin depth is smaller than the sample, the electrons are able to diffuse into or out of the skin depth in times comparable to the relaxation times and consequently a distortion of the ESR line appears (Dyson 1955).

This asymmetric e.s.r. line has been observed in metal–ammonia solutions by Catterall (1965), Chan, Austin, and Paez (1970) and Lelieur (1972). In each case, the asymmetry ratio (the ratio of the height of the low field maxima to the high-field minima for the derivative of the absorption line) has about the value expected if spin diffusion effects are negligible. In this case, the relaxation time T_2 is simply related to the line width ΔH. The e.s.r. line width in Na–NH$_3$ has been shown (Lelieur 1972) to increase from about 30 mG in dilute solutions to about 5 G at 10 m.p.m., while this line width is about 13 G near the melting point of pure liquid Na (Devine and Dupree 1970). Chan,

Austin, and Paez (1970) have shown that the electronic relaxation time T_2 is reduced in Cs–NH$_3$ relative to T_2 in Na–NH$_3$, which is due to the spin–orbit coupling.

2.2.7. Nuclear magnetic resonance

Duval, Rigny, and Lepoutre (1968) measured the Knight shift of ^{23}Na and ^{14}N in the Na–NH$_3$ solutions. One concentration is in the metallic range, the others in the intermediate range. Both shifts increase with concentration, and temperature. Their results indicate an equality of transverse and longitudinal relaxation times T_2 and T_1. Furthermore, the ^{23}Na relaxation time is of order 60 ms for a saturated Na–NH$_3$ which is much less than the value expected from the Korringa relationship (about 1 s).

Lelieur and Rigny (1973b) have measured the Knight shift of ^{133}Cs and ^{14}N in the concentrated Cs–NH$_3$ solutions. Their results for the variation of the ^{133}Cs Knight shift versus temperature are plotted in Fig. 2.13 for various Cs–NH$_3$ solutions. Using their magnetic susceptibility measurements of Na–NH$_3$ and Cs–NH$_3$ solutions, Lelieur and Rigny (1973b) have deduced the electronic density at ^{23}Na and ^{133}Cs versus the metal concentration (Fig. 2.14) and the temperature (Fig. 2.15). The ^{14}N Knight shift in

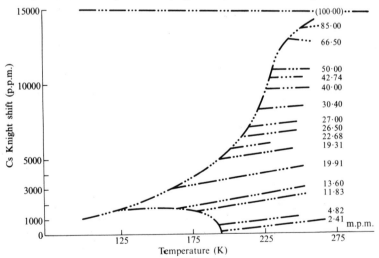

FIG. 2.13. The Knight shift at the Cs nucleus K(Cs) in Cs–NH$_3$ solutions. (Lelieur and Rigny 1973b). Concentrations are given at the right in m.p.m. The left-hand terminus of each line is fixed by the freezing line.

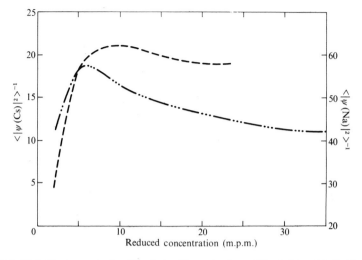

FIG. 2.14. Nuclear spin densities derived from the data of Figs 2.11 and 2.13. The concentration here is expressed in units of moles metal divided by moles metal plus moles NH_3 not in cation solvation layers. Solvation numbers of 4 and 8 are assumed for Na and Cs respectively.

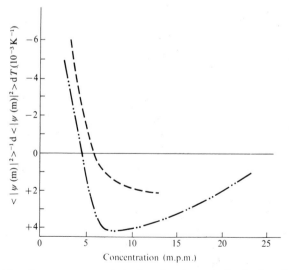

FIG. 2.15. Log temperature derivative of nuclear spin density from Fig. 2.14.

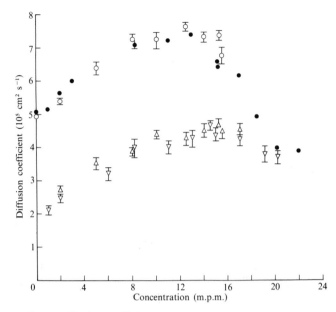

FIG. 2.16. The diffusion coefficient for Li, Na, and NH_3 in M–NH_3 solutions. (Garroway and Cotts 1973). The open circles are for ^{14}N in Na–NH_3 and the closed are for ^{14}N in Li–NH_3 solutions.

Na–NH_3 and Cs–NH_3 solutions have been found equal at the same concentration. The ^{133}Cs Knight shift has been measured versus the caesium concentration, even for concentrations greater than 20 m.p.m. At this concentration the ^{133}Cs Knight shift is about three times smaller than in the pure metal (Fig. 2.13). The electronic density at the metal nucleus displays different trends, being almost constant above 5 m.p.m. in Na–NH_3, while increasing in Cs–NH_3 solutions (Fig. 2.14). The differences in the solvation of Cs^+ and Na^+ ions are thought to be responsible for the differences in the electronic density at the metal nucleus.

More recently, Garroway and Cotts (1973) using the n.m.r. spin-echo, pulsed-magnetic-gradient technique, have measured the self-diffusion coefficients of ^7Li, ^{23}Na, and ^1H in Li–NH_3 solutions and in Na–NH_3 solutions. The lithium and ammonia self-diffusion coefficients have been found to increase with the metal concentration up to about 12 m.p.m. (see Fig. 2.16). For metal concentration greater than 12 m.p.m., the lithium self-diffusion coefficient levels off or perhaps decreases slightly, while the ammonia coefficient decreases by a factor of two, until it is essentially equal to the lithium self-diffusion coefficient at 20 m.p.m. Similar trends and values are obtained for Na–NH₃ solutions.

2.2.8. *Positron annihilation*

When an electron–positron pair annihilates, the annihilation gamma rays carry off the momentum of the pair (West 1973). If a positron is at thermal energy, then, the primary source of momentum in a metal is the electron with which the positron annihilates. Therefore, the angular correlation of annihilation gamma rays from positrons in a material indicates the electron momentum distribution in the material. Fig. 2.17 shows such an angular distribution for a lithium–ammonia solution (Varlashkin and Stewart 1966). The data thus far obtained (Varlashkin and Stewart 1966; McCormack and Millett 1966; Varlashkin 1968; Arias-Limonta and Varlashkin 1970) indicate that the distribution is altogether independent of concentration throughout the metallic range above 8 m.p.m. and furthermore that the momentum distribution does not change appreciably as one changes the

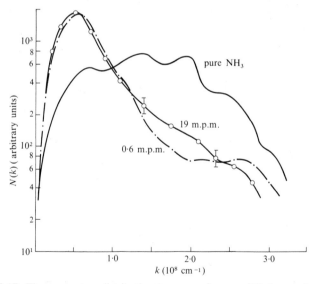

FIG. 2.17. The momentum distribution for γ-rays from annihilating positrons in Li–NH₃ solutions, and in pure NH₃. Data are shown for 0·6 m.p.m. and 19 m.p.m. solutions (Varlashkin and Stewart 1966).

solute ion under consideration. Nevertheless, the distribution is significantly different from that obtained in very dilute solutions or in pure ammonia (Holt, Chuang, Cooper, and Hogg 1968). The naive expectation that the distribution should show the parabolic shape appropriate to an assembly of free electrons is not, however, realized and the interpretation of these data is a controversial matter at present as is the presence or absence of the high momentum peak shown in the Fig. 2.17.

Morales and Millett (1967) have measured the lifetime (mean life) of positrons in lithium-ammonia solutions. The positron source, ^{22}NaCl, was also dissolved in the solution. The positron was found to be 'quenched' relative to pure ammonia, corroborating the early work of Hogg, Sutherland, Paul, and Hodgkins (1956). That is, the lifetime in the solutions ($0\cdot3 \times 10^{-9}$ s) was found to be much less than in pure liquid ammonia ($0\cdot8 \times 10^{-9}$ s), presumably as a consequence of the presence of the free electrons. The data are marred by the necessity of accounting for those annihilations which occurred in the glass walls of the container. The effect of the glass was determined by the somewhat unsatisfying procedure of subtracting a variable background fraction with the same shape as the distribution obtained with glass only. The background fraction (not shape) was adjusted so as to cause the data to lead to a single lifetime. The fraction subtracted was in the range of 10–20 per cent.

The magnetic data are qualitatively similar to those of simple liquid metals and do not require any unusual models for their interpretation. The positron data, on the other hand, are qualitatively different from simple metal data. It appears that the low electron density and the ammonia have an influence. We now turn from these specifically electronic data to a discussion of the available thermodynamic or mechanical data on the solutions.

2.2.9. *Density*

The density has been measured in solutions of lithium (Lo 1966), sodium (Kraus, Carney, and Johnson 1927; Kikuchi 1939; Naiditch, Paez, and Thompson 1967); potassium (Johnson and Meyer 1932; Demortier, *et al.* 1971); caesium (Hodgins 1949); calcium (Wa She Wong 1966). The data indicate that the density is a decreasing function of concentration approaching, in the

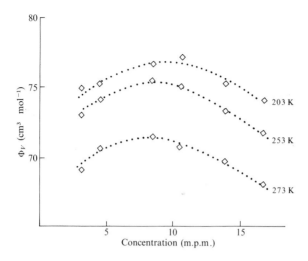

FIG. 2.18. Apparent molar volume Φ_V for K–NH$_3$ solutions at three temperatures (Demortier *et al.* 1971).

lithium solutions at least, a value smaller than that for any other liquid except the cryogenic fluids. The volume of the solution is approximately a linear function of the concentration of metal so that one may represent the data in terms of two parameters: one, the effective volume of the ammonia molecules in the solution V_A and the other, the effective volume of the metal in the solution V_M. This interpretation has long been applied to very dilute solutions of metals in ammonia. The second term (the apparent volume of metal per mole metal) is shown in Fig. 2.18 for K–NH$_3$ solutions. One sees that there is a slow decrease of this volume as the concentration is increased though the decrease is small and becomes smaller as the temperature is increased. Ichikawa and Thompson (1973) have also defined a partial molar volume as $\bar{v}_{Na} = \partial V/\partial n_{Na}$ where V is the solution volume and n_{Na} is the number of moles of metal at a given concentration. No real difference is obtained but a slight improvement in the thermodynamic foundation results. Alternatively, one may describe the metal ion and the ammonia in terms of hard spheres, consider the packing of these hard spheres, and extract a packing fraction η for the solutions. This latter approach will be discussed later when a theory is presented for the solutions.

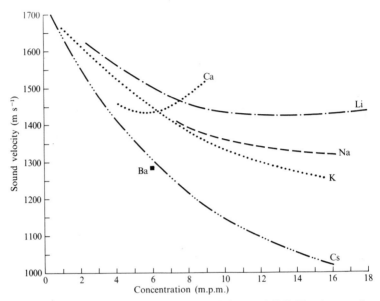

FIG. 2.19. Sound speed in several M–NH₃ solutions at 240 K. The data are due to Bowen and coworkers (Thompson 1973).

2.2.10. Compressibility

The velocity of sound has been measured in several metal-ammonia solutions (Maybury and Coulter 1951; Bowen, Thompson and Millett 1968; Bowen 1969; Bowen 1970; Thompson and Oré-Oré 1971; Bailey and Bowen 1972). It is a decreasing function of metal concentration, except near saturation (Bridges, Ingle, and Bowen 1970), as may be seen in Fig. 2.19, and a linearly decreasing function of temperature. The temperature coefficient of sound speed is shown in Fig. 3.27. The sound velocity and the density are combined to obtain the adiabatic compressibility of the solutions (see Fig. 2.2). Schindewolf and his co-workers (Schindewolf 1970; Boddeker and Vogelgesang 1971) have obtained crude values for the isothermal compressibility of the solutions, which are shown in Fig. 3.29 at zero pressure.

2.2.11. Heat capacity

Mammano in his study of solid Li(NH₃)₄, also measured the heat capacity of a saturated lithium-ammonia solution (Mammano and Coulter 1969). More recently Billaud (1972) measured the heat capacity of a 14 m.p.m. Na–NH₃ solution. These are the only available data on the heat capacity of M–NH₃ solutions. One finds (Fig. 7.3) that the heat capacity is a (more or less) linearly increasing function of temperature over the range from 90 to 200 K and that it increased by approximately 30 per cent over that range. The heat of solution has also been measured in these materials by several authors (Joannis 1889; Kraus 1908; Coulter and Monchick 1951) and recently by Gunn and Green (1962). In the metallic range of concentration the heat of solution is essentially independent of concentration. Absolute values are not available.

2.2.12. *Vapour pressure, viscosity, X-rays*

The vapour pressure of the concentrated solutions is small and approaches zero as saturation is approached, particularly in materials such as lithium–ammonia solutions (Marshall 1962). Further discussion is given in Section 2.2.13. The viscosity is also a decreasing function of concentration (Kikuchi 1944; Hutchison and O'Reilly 1970) and temperature (Demortier *et al.* 1971). There are no reliable X-ray data (Schmidt 1957; Brady and Varimbi 1964). In one case (Schmidt 1957) the data contain an undetermined contribution from decomposition products and, in the other, (Brady and Varimbi 1964) the data are restricted to small angles. More data are badly needed.

2.2.13. *Chemical potential and other thermodynamics*

Ichikawa and Thompson (1973) have reported extensive studies of electrochemical cells based on M–NH₃ solutions and chemical potentials derived from those measurements. Their e.m.f. measurements have been made using sodium β-alumina (Whittingham and Huggins 1971; Hsueh and Bennion 1971). This solid electrolyte removes the problem of finding an ionically conducting medium for Na^+ which is not attacked by, nor mixed with, either Na or Na–NH₃ solutions. Two types of cells without transference were employed.

cell A: Na (metal) | β-alumina crystal | Na–Hg amalgam

cell B: Na–NH₃ | β-alumina crystal | Na–Hg amalgam

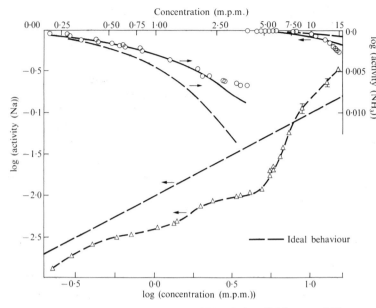

FIG. 2.20. Activity of Na and NH_3 in Na–NH_3 solutions (Ichikawa and Thompson 1973). Ideal behaviour is given by the long dashed lines. The solid line for a_{NH_3} represents data derived from vapour pressures (Marshall 1962).

The measured potentials E_A and E_B are related to the activity a_{Na} and/or chemical potential μ_{Na} in the mixture by:

$$E_A + E_B = -(RT/F)\ln a_{Na} = (\mu_0 - \mu_{Na})/F \qquad (2.3)$$

where F is the Faraday and μ_0 the chemical potential of pure Na. These chemical potentials together with the Gibbs–Duhem equation permit calculation of a_{NH_3}. Vapour pressure data (Marshall 1962) also may be used to calculate the same activity and, unfortunately, the two do not agree, as may be seen in Fig. 2.20. In each case the precision of the experiments is limited relative to more dilute solutions and the more accurate result cannot be determined. In either case there are pronounced departures from ideal behaviour at high metal concentration. One can immediately compute several other partial thermodynamic functions:

$$\Delta \bar{G}_{Na} = -F(E_A + E_B),$$
$$\Delta \bar{S}_{Na} = F[d(E_A + E_B)/dT],$$

and

$$\Delta \bar{H}_{Na} = F[T\{d(E_A + E_B)/dT\} - (E_A + E_B)],$$

where the left-hand sides, respectively, are the relative partial molar free energy, entropy and enthalpy of sodium in solution. As is usual with thermodynamics, one gains no additional information by this exercise but has a possibly more convenient presentation. Other analyses will be given below.

2.3. Comparison with other metallic liquids

All these properties of concentrated $M-NH_3$ solutions are qualitatively similar to those of simple liquid metals, (Cusack 1963; Allgaier 1969; Faber 1972) with the possible exception of the positron data, the thermopower, $d\sigma/dT$ and $d\sigma/dP$. The metal sodium may be taken as typical of simple liquid metals. Table 2.3, which is similar to one due to Egelstaff (1967), collects some of the data for liquid sodium near its melting point and for a

TABLE 2.3

Physical properties[a] *of simple liquids compared to* NH_3 *and* $Na-NH_3$ *solutions*

Property (units)	Ar	NH_3	Na–NH₃ (15 m.p.m.)	Na
1. m.p. (K)	84	195	180	371
2. b.p. (K)	87·5	240	—	1150
3. liquid range $= CP/TP$	1·8	2·1	—	7·5
4. density (gm/cm³)	1·4	0·72	0·62	0·93
5. heat capacity per atom or molecule				
C_p (erg/K)	5·06 k_B	8·8 k_B	6·4 k_B[b]	3.8 k_B
C_v (erg/K)	2·32 k_B	5·8 k_B	5·3 k_B[b]	3·4 k_B
6. compressibility isothermal $(10^{-11}\,cm^2/dyne)$	20	7·0	10	1·9
adiabatic $(10^{-11}\,cm^2/dyne)$	9	3·8	8·0	1·8
7. sound speed (m/s)	874	1720[c]	1320[c]	2500
8. viscosity (10^{-3} poise)	2·8	2·5	1·5	6·8
9. thermal conductivity (mW/cm K)	1·2	4·2	20	840

[a] At temperatures near the m.p. [b] Estimated for 15 m.p.m. $K-NH_3$ by Damay (1973). [c] At 240 K.

nearly saturated Na–NH₃ solution near 240 K. An extensive review of the properties of liquid metals has been given by Cusack (1963), and, more recently, the Brookhaven and Tokyo Conferences on Liquid Metals brought both theory and experiment up-to-date (Adams, Davies, and Epstein 1967; Takeuchi 1973).

In Allgaier's (1969, 1970) (Thompson and Allgaier 1970) scheme Na would exemplify Class A metals, while the M–NH₃ solutions more closely resemble Class B metals such as ZnSb. The distinction lies in both the magnitude of σ ($\sim 10^3 \, \Omega^{-1} \, \text{cm}^{-1}$) and the sign of $d\sigma/dT$. Class B metals generally have a Hall coefficient somewhat in excess of the free electron value, though the M–NH₃ solutions do not. It is this difference which keeps the solution data off Allgaier's (1970) universal curve of μ_H *versus* σ.

Though $d\sigma/dT$ is negative for liquid sodium, there are liquid metals e.g. Zn, Hg, which have $d\sigma/dT$ positive near their melting points. As will be shown below, this is a consequence of the way in which the local order of the metal changes with temperature relative to the characteristic electron wavelength, and is to be expected only for divalent metals. Metallic alloys e.g. Cd–In, sometimes show a reversal in the sign of the thermoelectric power. There is therefore, only the positron data for which there is no apparent analogue in the properties of simple metallic systems.

The density and Hall data indicate electron densities n in the range $1-4 \times 10^{21} \, \text{cm}^{-3}$, significantly lower than for the common metals. The r_s values, where

$$r_s = (\tfrac{3}{4}\pi n)^{\frac{1}{3}} a_0^{-1}$$

with a_0 the Bohr radius, are therefore above those for which the electron–electron interaction (correlation) may be safely ignored (Harrison 1970). This may be seen by comparing the Fermi and Coulomb energies of an electron at the average distance from another electron. In terms of r_s these energies are:

$$E_F = 3 \cdot 62 r_s^{-2} (Ry), \qquad (2.4)$$

and

$$E_C = 2 r_s^{-1} (Ry). \qquad (2.5)$$

The ratio E_C/E_F is near 3 for Cs, the least dense of the simple metals and near 1 for Al. The use, then, of ideal Fermi statistics in M–NH$_3$ solutions is open to question and the intrusion of correlation effects into the nearly free electron (NFE) theory must be expected. In the beginning, however, these will be ignored.

2.4. Ziman theory

The NFE theory (Ziman 1961, 1967, 1973; Bradley, Faber, Wilson, and Ziman 1962) as it will be used here, is based on two assumptions: (1) that the Fermi surface, separating filled from unfilled states in k-space, is spherical, and (2) that the density of states per unit energy $N(E)$ is that appropriate to a free electron gas, thus

$$N(E) = \frac{V}{2\pi}\left(\frac{2m}{\hbar^2}\right)^{\frac{3}{4}} E^{\frac{1}{2}}. \qquad (2.6)$$

In addition, the electron–ion interaction is to be treated in terms of the pseudopotential theory of Harrison (1966) and others (Phillips and Kleinman 1959; Heine and Abarenkov 1964; Ashcroft 1966). Finally, the mean free path Λ is to be obtained from the standard Boltzmann equation and is related to the conductivity by

$$\sigma = e^2 \Lambda \mathcal{S} / 12\pi^3 \hbar \qquad (2.7)$$

where $\mathcal{S}$ is the area of the Fermi surface. The relaxation time τ is then given by $\Lambda/\hbar k_\text{F}$. Mott and Jones (1936) have shown that

$$\tau^{-1} = \int d\mathcal{S}(1 - \cos\theta) P_{kk'} \qquad (2.8)$$

where the factor $(1 - \cos\theta)$ weights the effectiveness of a collision in reducing the current when an electron is scattered from a state K to a state K' such that K and K' are separated by an angle θ. The probability per unit time of scattering is

$$P_{KK'} = \frac{2\pi}{\hbar} |\langle K'|\, W\, |K\rangle|^2 \, N(E) \qquad (2.9)$$

according to Fermi's Golden Rule. $N(E)$ is the density of states, which is $h^2 k_F/m$ for free electrons. The matrix element has been shown by Ziman (1961) (Harrison 1966) to be factorable into a structure-dependent part and a structure-independent part if the scattering potential is assumed to be formed by a superposition of the potentials due to individual ions. The value is, if $\mathbf{k} = \mathbf{K} - \mathbf{K}'$:

$$|\langle \mathbf{K}'| \, W \, |\mathbf{K}\rangle|^2 = \frac{1}{V^2} |U(k)|^2 \left| \sum_i \exp(i\mathbf{k} \cdot \mathbf{r}_i) \right|^2 \qquad (2.10)$$

where V is the sample volume, $U(k)$ is the form factor, and the sum is over all the ions. Within the Born approximation, the form factor is

$$U(\mathbf{k}) = \int_V \mathrm{d}^3 r \exp(i\mathbf{k} \cdot \mathbf{r}) \, U(\mathbf{r}) \qquad (2.11)$$

with $U(r)$ the electron–ion potential. In most cases $U(r)$ will be a pseudopotential or model potential. The sum (in eqn 2.10) is related to the structure factor $a(k)$

$$a(k) = \frac{1}{N} \left| \sum_i \exp(i\mathbf{k} \cdot \mathbf{r}) \right|^2 \qquad (2.12)$$

The final expression for the resistivity ρ is then

$$\rho = \frac{4 k_\mathrm{F} m^2}{e^2 \hbar^3} \frac{n_\mathrm{s}}{n_\mathrm{e}} \int_0^1 \mathrm{d}y \, y^3 \, |u(2k_\mathrm{F}y)|^2 \, a(2k_\mathrm{F}y) \qquad (2.13)$$

where n_s is the number density of scatterers (ions), n_e the number density of electrons, and $y = k/2k_\mathrm{F}$.

The structure factor $a(k)$ can be obtained from X-ray or neutron diffraction data, though the experiments are difficult and the data sometimes unreliable. As an alternative, Ashcroft and Langreth (1967a, b) have undertaken to derive $a(k)$ from liquid state theory using the Percus–Yevick equation and treating the ion–ion interaction in terms of a hard-sphere potential. In the theory there are only a few adjustable parameters: the radius of the sphere and the packing fraction. Quite satisfactory agreement is found with the available data for simple metals. The extension of the calculation eqns (2.12) and (2.13) to a binary alloy (Faber and Ziman 1965; Ashcroft and Langreth 1967a, b) is straightforward, as the contributions of the three structure factors, connecting like (a_{11}, a_{22}) and unlike ions (a_{12}), are easily identified and computed.

The identification of the various contributions to the structure factor in an experiment requires both neutron and X-ray data on alloys of differing isotopic constitutions and is not easily accomplished (Enderby, North, and Egelstaff 1966).

The form factor $U(k)$ is ideally based on a knowledge of the crystal potential but is usually obtained from the pseudopotential formalism of Harrison (1966) or represented by a model potential (Ashcroft 1966) tailored to reproduce the resistivity or some other characteristic property. Several model potentials have been used:

$$U(r) = A\,\delta(r) \tag{2.14}$$

$$U(r) = -Br^{-1}\mathrm{e}^{-\lambda r} \tag{2.15}$$

$$U(r) = -\frac{Ze^2}{r\varepsilon(k)}\,S_R(r); \qquad S_R(r) = \begin{cases} 0,\ r < R \\ 1,\ r > R. \end{cases} \tag{2.16}$$

In each case, the effect of the electron–electron interaction is included through the use of the dielectric function $\varepsilon(k)$, usually the Lindhard (Hartree–Fock) function: (Lindhard 1954)

$$\varepsilon(k) = 1 + \frac{me^2}{\pi k_F \hbar^2 y^2}\left[\frac{1 - y^2}{4y}\ln\left|\frac{1+y}{1-y}\right| + \frac{1}{2}\right]. \tag{2.17}$$

The same formalism leads to two other properties: the isothermal compressibility K_T from $a(k)$ at $k = 0$ (Schroeder and Thompson 1969; Ashcroft and Langreth 1967a, b, 1968); and the thermoelectric power S. The latter is obtained by differentiation of the expression for the resistivity (Ziman 1961). The results are:

$$S = (\pi^2 k_B^2 T/3\,|e|\,E_F)[3 - 2q - r/2]; \tag{2.18}$$

and

$$K_T = (nk_B T)^{-1}a(0), \tag{2.19}$$

where

$$q = |U(2k_F)|^2\,a(2k_F)\left[\int_0^1 \mathrm{d}y\,y^3\,|U(y)|^2\,a(y)\right]^{-1} \tag{2.20}$$

and

$$r = 2E_F\left[\int_0^1 \mathrm{d}y\,y^3 a(y)\,\partial(|U(y)|^2)/\partial(y^2)\right]\left[\int_0^1 \mathrm{d}y\,y^3\,|U(y)|^2\,a(y)\right]^{-1}. \tag{2.21}$$

The quantity q comes from the explicit dependence of ρ on k_F, and r comes from the implicit dependence of the unscreened pseudopotential (hence also the model potentials) on the Fermi energy.

The application of these ideas to liquid metals has been highly successful. Ziman in his reviews (Ziman 1967, 1973) at the Brookhaven and Tokyo conferences on liquid metals could point to an understanding of how the positive sign for $d\sigma/dT$ could arise in divalent metals, and to success in computing many liquid metal parameters. The alloy theory of Ashcroft and others has been applied to several systems, including mercury amalgams with reasonable success (Evans 1970). One notable success was with the thermoelectric power of Ag–Au alloys (Howe and Enderby 1967) where a sign change is observed near 50%, and explained by Ziman's theory. The Hall coefficient follows the usual free electron form

$$R_H = (nec)^{-1} \qquad (2.22)$$

where c is the speed of light. Finally, the Wiedemann–Franz law is appropriate for any system in which the conduction is primarily electronic. In short, the NFE theory provides values for all of the low-frequency transport coefficients, and even K_T from $a(0)$ in close agreement with observation for nearly all liquid metals (Ascarelli 1968). Hg is the most prominent exception.

The optical constants have been considered an exception to the above but as experiment and theory improve, the agreement there also must be considered satisfactory (Smith 1969, 1970, Animalu 1967, Faber 1972).

Parameters such as the susceptibility can usually be computed from free electron (Fermi–Dirac) statistics. Properties, such as the positron-determined $N(k)$, appear to be beyond the theory at present though Ballentine (1966) has obtained $N(k)$ for Al, Zn, and Bi. Theories for the viscosity and similar mechanical parameters have not yet received much attention in metallic liquids (Cusack 1963, Faber 1972).

2.5. Qualitative theory of M–NH₃ solutions

Let us turn now to the application of these ideas to the solutions. There are several general points that can be made at once. Two independent sets of data indicate an effective electronic mass near the free electron mass and hence a wide band.

First, the magnetic susceptibility of a non-interacting free electron gas may be related to the effective mass m^* by:

$$\chi = \frac{n\mu_B^2}{2E_F}\left[3\frac{m^*}{m} - \frac{m}{m^*}\right] \tag{2.23}$$

where μ_B is a Bohr magneton. The results indicate m^*/m of order unity (Thompson 1965; Glaunsinger et al. 1972). Second, the optical value of the energy loss function may be related to the carrier density and the effective mass by a sum rule (Pines 1962):

$$n = \frac{\varepsilon_\infty m^*}{2\pi^2 e^2}\int_0^\infty \text{Im}[\varepsilon^{-1}(\omega)]\omega\,\mathrm{d}\omega \tag{2.24}$$

where ε_∞ is the result of the background polarization. The sum rule is independent of the Drude theory and is valid in the absence of interband transitions (virtual or real) (Faber 1966). This sum rule is to be preferred in the case of metal–ammonia solutions to the more conventional one:

$$n = \frac{\varepsilon_\infty m^*}{2\pi^2 e^2}\int_0^\infty \text{Im}[\varepsilon(\omega)]\omega\,\mathrm{d}\omega \tag{2.25}$$

which requires a knowledge of $\varepsilon(\omega)$ when $\omega \gg \omega_p$. The solutions, because of their low electron density, have $\omega \sim \omega_p$. These integrals yield m^*/m values again near unity, though not identical to those derived from χ.

The resistivity (or conductivity) as given by eqn (2.13) is nominally a function of k_F. There is, however, no universal $\rho(k_F)$ present in these solutions (Fig. 2.21). The conductivity parameter

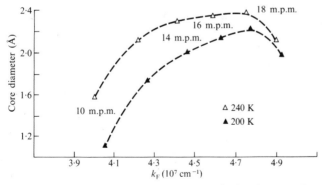

FIG. 2.21. Required diameter for empty-core potential for electron-solvated ion interaction (Schroeder and Thompson 1968; Thompson 1973).

(see § 2.6.2)' at two temperatures is plotted against k_F. Changing k_F at fixed T by varying the concentration does not have the same effect on σ as changing k_F at fixed x by varying T. Varying k_F by varying the pressure produces an effect similar to that produced by changing T. There is a weak dependence of σ on solute ion; in the alkali metal solutions σ decreases as we go from Li to Cs (except for Rb–NH$_3$ solutions which have the lowest σ) but not by so large a fraction as in the pure metals. The conductivity in the pure metals is not a monotonic function of atomic weight. Also the mean free path is long enough that the pseudopotential approach seems justified. The electron wavelength is great enough that details of the potential may not be required.

2.6. A specific model: solvated ion scattering

The present model, in which the solutions are assumed to be a mixture of solvated metal ions, of ammonia molecules, and of free electrons, is based on a suggestion of Maybury and Coulter (1951). It has been used by Schroeder and Thompson (1969; Thompson 1973) and has recently received strong experimental support (Garroway and Cotts 1973). The analysis of the experimental values for the self-diffusion coefficient D (Garroway and Cotts 1973) show that the diffusing species do not appear to be free metal ions and free ammonia molecules. In fact, the smaller lithium (or sodium) self diffusion coefficient indicates that the metal ion is impeded in its motion. For a saturated Li–NH$_3$ solution, the self-diffusion coefficients of Li[7] and H[1] are practically equal, which is a strong indication that the lithium ions are solvated by four ammonia molecules. At lower concentrations, the self-diffusion coefficients for Li[7] and H[1] are different and a solvation number cannot be unambiguously determined, but the experimental values of these coefficients imply some solvation of the lithium ion. The Na–NH$_3$ system saturates at 15·5 m.p.m., and the Na[23] and H[1] self-diffusion coefficients are not equal at this concentration. Therefore the solvation number of the sodium ion cannot be exactly prescribed; but can only be shown to be smaller than 5·45, and the strong similarity of Li–NH$_3$ and Na–NH$_3$ diffusion data has suggested to Garroway and Cotts that the sodium ion is also solvated by four ammonia molecules.

Using the Stokes–Einstein expression for D, and the experimental viscosity, the calculated radii for the $Li(NH_3)_4^+$ and $Na(NH_3)_4^+$ complexes have been found essentially equal (approximately twice as large as the free ammonia radius), and concentration independent, indicating that the solvation number is not strongly concentration dependent.

There are two stages in using this model for the calculation of physical properties.

2.6.1. *The structure factor*

First, a discussion is given of the interaction among the solvated ions and ammonias and the calculation of $a_{ij}(k)$. Second, a description is given of the electron–ion and electron–molecule potentials and the calculation of $U_i(k)$. From these σ is calculated, as is the thermopower. The first stage of the computation receives a partial check in that the compressibility K_T may be obtained from $a_{ij}(0)$. However, the resistivity and thermopower depend primarily on $a_{ij}(2k_F)$, which may be inappropriate for their calculation even if the structure factor at $k = 0$ yields the observed K_T. The test of the model must then be based on a calculation of more than K_T. We turn now to the structure factor.

The structure factors $a_{ij}(k)$ are related to the Fourier transform of the radial distribution functions, $g_{ij}(r)$. The radial distribution function, $g_{11}(r)$, is defined such that $n_1 g_{11}(r) 4\pi r^2\, dr$ gives the number of type 1 particles in a spherical shell of radius r centered about a type 1 particle; n_1 is the average density of type 1 particles. g_{22} refers to type 2 particles with respect to type 2, and g_{12} ($= g_{21}$) refers to type 2 particles with respect to type 1. Naturally for a one-component system there is only one structure factor.

For a one-component system

$$a(\mathbf{k}) = \frac{1}{N}\left|\sum_i e^{i\mathbf{k}\cdot\mathbf{r}_i}\right|^2 \qquad (2.26)$$

where the sum extends over all N ions. In the liquid a thermal average of the r_i is required. In the present case we have assumed a binary mixture and, following Ashcroft, (Ashcroft and Lekner 1966; Ashcroft and Langreth 1967a, b) we assume the massive constituents (solvated ions and unbound NH_3 molecules) to interact as hard spheres. Ashcroft makes use of a treatment of a hard-sphere liquid by Lebowitz (1964) using Percus–Yevick theory; they find the structure factors to be completely determined by four interrelated parameters: (1) the radius of component A, R_A; (2) the radius of component B, R_B (taken as the larger, i.e., $R_A < R_B$); (3) the packing fraction η which is the ratio of the total volume of the two kinds of spheres to the volume of the system; and (4) the fraction x' of constituents A:

$$x' = n_A/(n_A + n_B),$$
$$= n_A/n \qquad (2.27)$$

where n_A and n_B are the number densities of spheres of types A and B, respectively. Only the ratio $\alpha = R_B/R_A$, and x' need be known to fix $a_{AA}(k)$, $a_{BB}(k)$, and $a_{AB}(k)$. The structure factors are then evaluated at $k = 0$ to obtain K_T. The advantage of the Lebowitz formulation is that all the structure factors are given by closed, algebraic formulae. For example, K_T is given by

$$nk_B T K_T = (1-\eta)^4 \{(1+2\eta)^2 - \Delta\}^{-1} \qquad (2.28)$$

where

$$\Delta = \frac{3x'(1-x')\eta(1-\alpha)^2}{x'+(1-x')\alpha^3}\bigg\{(2+\eta)(1+\alpha) + \\ + \frac{3\eta\alpha}{x'+(1-x')\alpha^3}\{(1-x')\alpha^2 + x'\}\bigg]. \qquad (2.29)$$

For a one-component system $x' = 0$ and Δ vanishes.

The values of R_A, R_B, α, η, and x' for the solutions were determined by Schroeder and Thompson (1969) as follows. Li–NH$_3$ solutions were considered first. Let component A be the unbound ammonia molecules and B the solvated ion. Pure liquid ammonia is treated as a one-component (A) fluid and η is obtained from K_T (eqn (2.28) with $\Delta = 0$). Then R_A is found from $\eta = 4\pi R_A^3 n_A/3$ with n_A obtained from the known density of pure NH$_3$. Typical values of R_A lie near 1·5 Å.

The description of a solvated ion, component B, required a specification of the number z of solvent molecules bound to the ion. It is concluded from the existence of Li(NH$_3$)$_4$ (see Chapter 7), and from the diffusion constant (Garroway and Cotts 1973), that *four* is an appropriate co-ordination number. Hence, a 20 m.p.m. Li–NH$_3$ solution is also a one component system as there are no free ammonias. Therefore one may expect to obtain R_B from the isothermal compressibility of a 20 m.p.m. Li–NH$_3$

solution. However, K_T is not known for Li–NH$_3$ solutions and we must invoke the principle that all solutes are much alike. The isothermal compressibility data due to Schindewolf (1970) and Böddeker and Vogelsgesang (1971) are combined with the general observation that $K_T = K_S$ in metallic liquids to conclude that we might use the adiabatic compressibility K_S in lieu of K_T. Indeed K_T (15 m.p.m. K–NH$_3$ at 240 K) $= 95 \times 10^{-12}$ cm^2 dyn^{-1} (Schindewolf 1970), and K_S (15 m.p.m. Li–NH$_3$ at 240 K) $= 93 \times 10^{-12}$ cm^2 dyn^{-1}, (Bowen $et\ al.$ 1968), as hoped. Thus R_B may be determined. Typical values lie near 3 Å.

At any given intermediate concentration x, α is known ($= 2{\cdot}0$) and x' may be determined from x and the coordination number z ($= 4$). Finally, η is obtained from R_A, R_B, and the known density. The entire calculation is based on the known properties of the pure solvent, the saturated solution, and the density as a function of x and T. No parameters are adjusted for $0 < x < 0{\cdot}20$. The substitution of K_S for K_T is particularly distressing when a model is being tested; there is no choice. As stated, K_T and K_S are quite close in most metallic liquids, typically differing by no more than 15 per cent. Even in insulating liquids the ratio seldom exceeds $1{\cdot}5$. In any case the proper reproduction of the temperature and concentration dependence of K provides a more severe test than simply obtaining the proper magnitude.

The results of the calculation are shown in Fig. 2.22 where

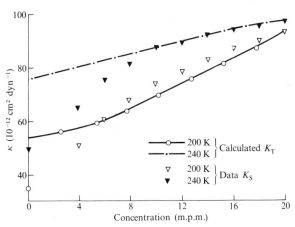

FIG. 2.22. Compressibility of Li–NH$_3$ solutions. The data are depicted by closed triangles for 240 K and open for 200 K. The lines are calculated from $a(0)$ (Schroeder and Thompson 1969).

the data (Bowen, Thompson and Millett 1968) for K_S are given by the circles and squares and the lines represent the result of the calculation. It is clear that the theory has reproduced both the concentration and temperature dependence of the compressibility. The success is marred somewhat by the necessity of substituting K_S for K_T but must be taken as real nonetheless. The accuracy of $a_{ij}(0)$ is thus assured and the extension of the calculation of a_{ij}s to nonzero k may be made with some confidence.

Though X-ray data would provide a test of $a(k)$ at finite k, the available solution data cannot be compared to the calculation (Schmidt 1957, Brady and Varimbi 1964). Furthermore, the coherent contributions of the molecular structure to the diffraction pattern from polyatomic liquids are not easily removed, as has been recently discussed by Pings and Waser (1968). The calculated $a(k)$ for pure liquid NH_3 nevertheless bears a strong resemblance to Narten's data, (Narten 1968) Fig. 2.23. In that case, at least, the calculation has properly reproduced the main features of the data. Typical values of $2k_F$ are near $0.8 \times 10^8 \, \mathrm{cm}^{-1}$, well to the left of the main peak.

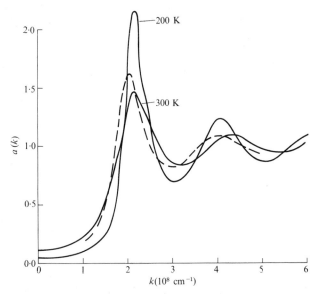

FIG. 2.23. Structure factors $a(k)$ of NH_3 at 200 K and 300 K. The dashed line shows a modified structure factor at 277 K due to Narten (1968; Schroeder and Thompson 1969).

Before going on to the calculation of σ let us look over the character and consequence of the present results. It must first be noted that several alternative calculations yield neither the x nor T dependence of K_T: the free-electron gas equation which relates the sound speed to the Fermi velocity (the Fermi velocity is low); the calculation of eqns (2.28) and (2.29) with an ion co-ordination number of zero, two, or six. The choice of $Li(NH_3)_4^+$ as the ion is thus necessary to the success of the calculation, as is the inclusion of the hard sphere interaction. The values of R_A and R_B are found to vary slightly with T; no defence can be offered for this variation other than the extreme simplification of the model. Similar variations were found by Ashcroft in his calculations for much simpler liquid metals and alloys. Similar calculations for pure Rb have been reported and suffer similar shortcomings (Jarzynski, Smirnov, and Davis 1969). The magnitude of R_A ($\sim 1 \cdot 5$ Å) fits estimates of the size of the NH_3 molecule based on gas viscosity and van der Waals parameters. R_B is about twice R_A as might be expected from the structure assumed for the solvated ion. The values of the packing fraction η ($\approx 0 \cdot 5$) are slightly larger than those found for the simple metals ($\approx 0 \cdot 45$). This may be attributed to the assumption of a spherical shape for the solvated ion. The actual tetrahedral shape allows for better packing. It may alternatively derive from the inequality in sphere radii. The low η ($\approx 0 \cdot 35$) of the pure solvent reflects the open structure characteristic of polar liquids as well as the oversimplification of the intermolecular interaction in the hard-sphere model.

There is, however, another, deeper, problem: the volume of the solutions greatly exceeds that of the constituents. This and other reasons have led to the development of a model for the dilute solutions based on the concept of an electron-in-a-cavity (see Chapter 3). The metal valence electron is presumed to exist in a bubble in the liquid of radius near $1 \cdot 7$ Å. Almost certainly the cavity does not exist throughout all of the metallic range of concentrations. The volume is thus explained at one extreme of concentration in terms of the packing of solvated ions and ammonia molecules while at the other, one introduces a third, very large constituent. Yet, and this is the problem, the volume is a smooth function of x and may be represented by a function of the form

$$V_S = V_A + x V_M \tag{2.30}$$

where V_S is the solution volume, and V_A and V_M represent effective volumes of ammonia and metal atom (or metal ion plus electron), respectively. It would seem that there should be some wiggle in the curve of V_S versus x when the cavities disappear, and that V_M should be different in dilute and in concentrated solutions. Yet in Na–NH$_3$ solutions and K–NH$_3$ solutions V_M varies by only 10 per cent over four decades of concentration. Data do not exist elsewhere. This point will be taken up again in Chapter IV and a plausible, yet not completely convincing, resolution obtained.

Calculations of compressibilities for Na, K, Cs, and Ca dissolved in liquid ammonia have been made (Thompson 1971). Since a composition equivalent to Li(NH$_3$)$_4$ cannot be ascertained, the parameters of the model were fixed by the following scheme. The free ammonia radius R_A was fixed to be ≈ 1.5 Å from pure NH$_3$ compressibilities as before. The solvated ion radius R_B was then varied for several choices of the co-ordination number z. R_B and z were chosen so as to give the best fit to the data while not deviating from the range of values established for the Li–NH$_3$ system.

The fit was considered successful when for a single value of z, the spread in the values of R_B required for an exact fit was minimal. This process yielded both a value for z and an (average) value for R_B. The average of the differences between calculated and measured values was less than 2 per cent for each solute. Table 2.4 gives the values of the parameters used. In each case the choice of co-ordination number z confirmed prejudices based on chemical experience with alkali ions in various solvents (Evans 1964). Calculations were necessarily inaccurate when there were too few ammonia molecules to supply full co-ordination (as in the higher concentration Cs–NH$_3$ solutions). This model provides only a parametrization of the measured compressibilities and should not be relied on for an accurate representation of the structure of the solutions. The parameters used are, nevertheless, eminently reasonable. The ionic sizes used in the model bear, for example, a nearly constant ratio to the sizes expected for solvated ions. These results lend considerable support to the model which considers concentrated metal–ammonia solutions to be mixtures of solvated ions and free ammonia molecules and which treats their interaction in a hard-sphere approximation.

TABLE 2.4
Parameters of hard-sphere model of M–NH₃ solutions

Solute	Co-ordination number	Diameter of solvated ion (Å)	Low and high values of packing fraction η
Li	4	6·30	0·42–0·54
Na	4	6·45	0·43–0·49
K	6	7·33	0·47–0·58
Cs	8	8·04	0·47–0·62
Ca	6	7·43	0·43–0·52

2.6.2. The form factor

We turn now to the second factor in the integral of eqn 2.13: $U(k)$, the form factor. A calculation of even the pseudopotential form factor is beyond the present capabilities of the theory. One must therefore search for plausible and workable approximations to the potentials of both the solvated ion and the ammonia molecule.

Pohler and Thompson (1964) used a screened Coulomb potential for the ions, ignored the ammonia, took $a(k) = 1·0$, and nevertheless obtained a resistivity at saturation within an order of magnitude of the observed value. Neither the concentration nor temperature dependence could be reproduced. Schroeder and Thompson (1973) used a repulsive δ-function approximation to layer to the screened Coulomb potential of the ion and could obtain a better fit because of the large number of adjustable parameters. Neither calculation included the effects of the structure factor so the quality of the potential could not be judged. Ashcroft and Russakoff (1970) ignored the repulsive part of the pseudopotential but considered the effect of the angular dependence of the dipole field. Their fit was excellent at the one temperature considered. Schroeder and Thompson (1968) Thompson (1973) used a repulsive δ-function approximation to the pseudopotential, ignored the dipole field and obtained results equivalent to Ashcroft and Russakoff.

Ashcroft and Russakoff (1970) took advantage of the cancellation which occurs when random dipole potential terms are treated in first Born approximation. They argued that the efficacy of electron screening is such as to reduce the tendency of dipoles to align so that angular orientations are independent of spatial configurations. The sums which led to dipole-dipole structure factors for spherically symmetric potentials then lead to independent scattering so that the dipole–dipole structure factor is independent of k and the part of the resistivity due to scattering of electrons by dipoles is proportional only to the number of free dipoles N_d. Cross terms involving charge–dipole structure average to zero on the same basis. Only the ion–ion structure factor a_{ii} remains and was calculated as by Schroeder and Thompson (1968). They adjusted the moment of the bound NH_3 dipole to 125 per cent of the free dipole moment to account for the polarization produced by the field of the ion. One has then that the resistivity is proportional to

$$N_d \int_0^1 U_d^2(y) y^3 \, dy + N_i \int_0^1 U_{si}^2(y) a_{ii}(y) y^3 \, dy \qquad (2.31)$$

where U_d is the point dipole potential and N_d, N_i are the number densities of free dipoles and solvated ions, respectively. In eqn (2.31) the solvated ion potential U_{si} includes the Coulomb field of the ion together with the dipole fields of the solvating NH_3 molecules. The dipole contributions to U_{si} tend to cancel the ionic term and account for the long mean free path in saturated solutions.

Schroeder and Thompson argued that the high dielectric constant of the ammonia together with the free electron screening make any long ranged potential unlikely. The dipole potential, in any event, will be much weaker than the Coulomb. It seems more likely that the repulsion from filled orbitals will dominate. They therefore chose as a first approximation the simplest form factors for the ammonia and solvated ion potentials: constant form factors which are equivalent to delta function potentials. The choice of values of U_A and of U_B, the form factors of solvent molecule and solvated ion respectively, was made empirically, as follows. One again treats a 20 m.p.m. Li–NH_3 solution as a single component fluid. In such a system, with $U(k)$ independent of k we have $\sigma^{-1} \propto |U_B|^2$. Thus U_B can be fixed, except for sign. A value near ± 0.4, in units of $\frac{2}{3} E_F$, is obtained.

Schroeder and Thompson then computed $\sigma(x, U_A)$ and adjusted U_A to fit the data. There is again an ambiguity in sign. Only the relative sign of U_A and U_B is important, as only the product U_A and U_B enters eqn (2.13a) without an absolute value sign.

If U_A and U_B are to be independent of x, the two potentials must be chosen of like sign. There thus results an unphysical divergence of σ for some U_A near 0·1. Yet for $U_A = 0·4$, in units of $\frac{2}{3}E_F$, the agreement extends throughout the metallic concentration range (8 to 20 m.p.m.) at 240 K. The only adjustable parameter is U_A.

The agreement in either calculation must be regarded as exceedingly good, especially when the wide range of conductivities is considered. The immediate implication is that the large change of σ with x is primarily a consequence of the change in local order, i.e. $a(k)$, with x, and of the decrease in the free ammonia fraction, not of the change of k_F. Before explaining further consequences of the models they are applied to a determination of $d\sigma/dT$.

2.6.3. $d\sigma/dT$ and S

An increase of T increases $a(k)$ while decreasing k_F, in a one component monovalent system. One thus expects an increase of σ with T, if $a(k)$ dominates. However, the decrease in k_F is the more important effect. Thus $d\sigma/dT < 0$. Only by taking dU_B/dT as negative can $d\sigma$ be made positive in the Schroeder–Thompson model. In a 20 m.p.m. solution the required change in U_B is $-0·18$ per cent per degree. We again adjust U_A at lower concentrations and find U_B to change by $-0·19$ per cent per degree. Ashcroft and Russakoff have not reported an attempt to calculate $\sigma(T)$. It is clear that they also must adjust the potential, probably by (1) decreasing the dielectric constant and (2) enlarging the solvated ion as T increases.

An explanation of $d\sigma/dT > 0$ can be gained only by the ad hoc assumption of a temperature dependent potential (Catterall and Mott 1969).

The Ashcroft normalization of $U(k)$ in eqn (2.13) (units of $\frac{2}{3}E_F$) already inserts some temperature dependence through the change of E_F with temperature (i.e. density). This amounts to about $+0·05$ per cent per degree. The net change in the potential is thence near $-0·15$ per cent per degree. Similar problems have been faced by others when using model potentials. Such changes appear to result from the over-simplification of a complex problem, and from the energy dependence inherent in the pseudopotential technique. In the present case there is also the possibility that the ammonias bound to the ion are less tightly bound as T increases. This would reduce the electron density associated with the solvated ion and perhaps reduce the repulsion. No such reduction can be expected for an isolated NH_3 molecule, of course. See Thompson (1973).

Catterall and Mott (1969) point out the possibility that it is the ammonia relative permittivity which produces the increase in σ with increasing T. They argue that the ion–ion correlation is governed by the static dielectric constant of the ammonia which drops off very rapidly with an increase in T. Thus $a(k)$ would also drop (this supposes a more accurate calculation of $a(k)$ than the hard-sphere model given above) and σ would increase. Only the small number of NH_3 molecules not involved in solvation would be expected to contribute to this screening, and the value of $d\sigma/dT$ would then approach that of simple metals at higher concentrations (as it does).

It must be noted that the representation of the form factor by a temperature dependent constant in the Schroeder–Thompson model has successfully reproduced the observed σ, $d\sigma/dx$ and $d\sigma/dT$. Let us now look at the values of U_A and U_B required for this success.

First, U_A and U_B both have the same sign. We have implied above that U_A is primarily derived from the repulsion of the electron by its exchange interaction with the ammonia molecule. Thus U_A is expected to be positive. Why should U_B be positive? It must be that the electron and dipole screening largely obscure the Coulomb field of the ion, leaving only the core repulsion. Remembering that the 'ion' includes several NH_3 molecules one must conclude that the Coulomb attraction compensates for about three-quarters of the core repulsion, leaving $U_B \sim U_A$. This allegation can be verified only by a complete pseudopotential calculation.

Consider now the thermoelectric power parameter q eqn (2.20). It depends primarily upon $a(2k_F)|U(2k_F)|^2$, or the analogous binary combination. The observed values (for Na–NH_3 solutions) all lie near 1·5. The computed values (Schroeder and Thompson 1968) cover an order of magnitude from 2 to 20.

There is no agreement between theory and experiment. But as Robinson (1967), Robinson and Dow (1968), and Meyer and Young (1969) have argued, the thermopower is sensitive to the pseudopotential, particularly the energy dependence. Terms involving $(\partial U/\partial E)_{E=E_F}$ appear and can be as large as q. The proper treatment of the influence of the energy dependence of the mean free path on the thermopower has only recently been given. With the highly simplified potential used here one could not expect close agreement.

2.6.4. *Summary of model*

Each of these calculations is consistent with the observation that it is the increasing order of the solutions and the decreasing NH_3 fraction with increasing metal concentration which causes the increase in conductivity. Each also requires a weakening of the potential as T increases. The success of the calculation leads to the conclusion that metal–ammonia solutions containing more than 8 mole per cent metal may be described as a binary mixture of solvated metal ions and solvent molecules, permeated by a free electron gas.

The 'dressing' of each ion by a sheath of ammonia molecules is doubtless responsible for the slight dependence of the conductivity, among other parameters, on the particular solute considered.

Several previous investigators, led astray by the positive temperature coefficient of conductivity, have attempted to describe the solutions as liquid semiconductors. Arnold and Patterson (1964) suggested that conduction occurred in an 'excited state band.' They determined an activation energy from $d\sigma/dT$, deduced a carrier density, then used a screened Coulomb potential to calculate $\sigma(x)$. They obtained quite good agreement with the observed σ but their carrier densities could not be reconciled with the Hall effect data. They ignored any structure effects as well as scattering by the ammonia. This theory must be rejected in favor of the NFE. The NFE theory immediately provides a qualitative understanding of such other transport properties as the thermal conductivity, the Hall coefficient, and the optical constants.

Only the latter requires more than the conventional electron gas treatment. Measurements of the optical constants of solid metals form the basis of a standard technique in research on the band structure of metals, as Cooper, Ehrenreich, and Philipp

(1965) have shown in their work with solid copper, silver, and gold. Butcher (1951) examined the alkali metals, customarily thought to be free-electron metals, and found an excess optical absorption at low energies that he attributed to interband transitions. The influence of band structure on the optical properties of solid metals is in contrast to the behaviour of the pure liquid metals, where the optical properties may be adequately described by a free electron calculation based on the d.c. conductivity and the valence-electron density. This apparent simplicity may be a consequence of the limited energy range of the reported work though no interband transitions have been found in liquid Hg at energies up to 20 eV (Bloch and Rice 1969) in contrast to the situation in solid Hg (Mueller and Thompson 1969). The success of the conductivity calculation leads to the expectation that the properties of the solutions be derivable from a theory similar to those Faber (1966, 1972) in particular, has applied to the optical properties of liquid metals.

2.7. Deductions from optical data

We will not, however, present a computation of the optical or other properties of metal-ammonia solutions, rather interpret the results in the spirit of the above theory.

In contrast to the simpler metals with a negligible core polarization contribution to the relative permittivity, metal–ammonia solutions have a background polarization due to the NH_3 molecule. Thus the high frequency limit of the relative permittivity ε_∞ is almost twice the vacuum value. There are several consequences of this result. The plasma frequency is shifted downward from the usual value by $\varepsilon_\infty^{-\frac{1}{2}}$; the peak in the energy loss function is also shifted downward by $\varepsilon_\infty^{-\frac{1}{2}}$; the values of the sum rule integrals are shifted as well. It is necessary to estimate an effective value of ε_∞ as the ammonia molecules occupy only a fraction of the volume of a sample. This has been done by the following simple device. The optical relative permittivity of NH_3 is 1·9. One assumes this value obtains in the volume fraction occupied by NH_3, and the vacuum value (unity) to obtain elsewhere. The effective relative permittivity (in the limit of infinite frequency) is then given by dividing the relative permittivity among the constituents according to their volume (Vanderhoff *et al.* 1971):

$$\varepsilon = 1\cdot 9\alpha + (1 - \alpha) \tag{2.32}$$

where α is the volume fraction of ammonia molecules in the bulk.

Next consider the sum rules, as they are independent of the model chosen to describe the conduction process. Two forms of the sum rule (Ehrenreich, 1966b) relate $\mathrm{Im}[\varepsilon(\omega)]$ or $\mathrm{Im}[\varepsilon^{-1}(\omega)]$ to the conduction electron density:

$$n = \frac{\varepsilon_\infty m^*}{2\pi^2 e^2} \int_0^\infty \mathrm{Im}[\varepsilon(\omega)]\omega \, d\omega, \tag{2.33}$$

and

$$n = \frac{\varepsilon_\infty m^*}{2\pi^2 e^2} \int_0^\infty \mathrm{Im}[\varepsilon^{-1}(\omega)]\omega \, d\omega. \tag{2.34}$$

In writings eqns (2.33) and (2.34) the plasma frequency is assumed to be given by

$$\omega_{\mathrm{p}}^2 = 4\pi n e^2 / m^* \varepsilon_\infty. \tag{2.35}$$

Note that the effective medium relative permittivity ε_∞ tends to increase the value obtained from eqn (2.33) while reducing that from eqn (2.34). One thus expects the values of n derived from eqns (2.33) and (2.34), respectively, to have a ratio of ε_∞^2. As $\mathrm{Im}[\varepsilon^{-1}(\omega)]$ is small outside the range of available data, eqn (2.34) provides the more reliable estimate of n. An extrapolation of $\omega\varepsilon_2$ to zero frequency cannot be made unambiguously. The Hall data is used as a guide to the extrapolation and not eqn (2.33) to determine n.

Table 2.2 contains values of n determined from eqn (2.34) as well as the relevant Hall data. In evaluating eqn (2.34) the effective mass ratio is unity. Clearly the optical and Hall data do not completely agree. Faber (1966) has shown that the sum rule may yield values of n in excess of the free electron value as a consequence of excitation of core electrons. In the case of Li such excitation would not be likely to contribute much before 4 eV. (Mueller and Thompson 1969, Vanderhoff *et al.* 1971.) More important in the present case is the fact that the sum rule result for n lies so near the free electron value. The slight difference could easily be removed by introducing an effective mass $m^* \neq m$.

The question of an effective mass will be taken up in more detail below, following the analysis using the Drude model. The present agreement between sum rule and Hall effect determinations of the electron density is as close as might be expected.

The medium relative permittivity, whether it derives from ion core polarization as in simple metals, or from the polarization of the ammonia molecule as in metal–ammonia solutions, operates on the usual Drude equations as a simple multiplier (Drude 1900, Stern 1963):

$$\varepsilon_1 = \varepsilon_\infty[1 - \omega_p^2\tau^2/(1 + \omega^2\tau^2)], \qquad (2.36)$$

and

$$\varepsilon_2 = \varepsilon_\infty\omega_p^2\tau/[\omega(1 + \omega^2\tau^2)]. \qquad (2.37)$$

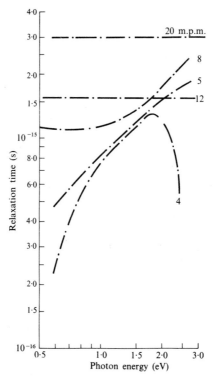

FIG. 2.24. Optical relaxation times as calculated from Drude fits to the measured relative permittivity for several Li–NH$_3$ solutions (Thompson and Cronenwett 1967).

In eqns (2.36) and (2.37), τ is the relaxation time. The results of calculations based on the Drude model and d.c. parameters also are shown in Fig. 2.10. In the calculation, values of ε_∞ taken from eqn (2.32) were used. Again, there is incomplete agreement between theory and experiment. In fact, the ε_1 data for the 4 and 5 m.p.m. solutions show so little resemblance to the Drude form that we will not attempt further analysis of those data on a simple Drude basis but defer their discussion until later. By combining eqns (2.36) and (2.37) one can obtain optical values for τ and ω_p. These values, to be referred to as τ_D and ω_p^D, are shown in Figs 2.24 and 2.25. Though both τ_D and ω_p^D are affected by the choice of ε_∞, n/m^* from eqn (2.35) is independent of ε_∞. A common criterion for the applicability of the Drude model is that τ_D and ω_p^D be independent of photon energy, clearly both depend slightly on $\hbar\omega$ in the present case. However, as may be seen in Table 2.4 frequency-averaged values of τ and ω_p are not very different from values of these parameters derived from d.c. data, except for the relaxation time for the 20 m.p.m. solution.

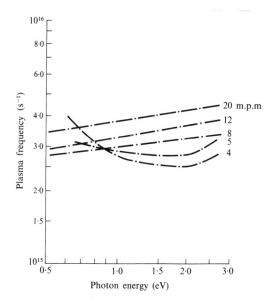

FIG. 2.25. Plasma frequencies as calculated from Drude fits to the measured relative permittivity for several Li–NH₃ solutions (Thompson and Cronenwett 1967).

Wilson and Rice (1966) and also Faber in his 1966 paper pointed out that, when the photon energy is of the order of the Fermi energy, as in the present work, the usual Drude theory must be amended. The basis of the required change is the large energy given to the electron which removes it from the vicinity of the Fermi surface. The response of the electron then depends on matrix elements involving states well above the Fermi energy. Unfortunately the expected deviations from the Drude theory are complicated functions of the photon energy and of the pseudopotential matrix elements. The application of these ideas to the present data is difficult. One may only observe that the present data show trends similar to those computed by Wilson and Rice and by Faber. A more detailed investigation is required of both theory and experiment before meaningful comparisons can be made.

Two other papers have discussed the consequences of nonuniform surface density for the interpretation of optical data. Bloch and Rice (1969) point out that the charge and mass densities do not drop discontinuously from bulk to vacuum values (or sample to window values) rather change smoothly as the surface is approached. The intensity and polarization of reflected light is thus not determined altogether by the properties of the bulk. Ellipsometric measurements prove to be particularly affected by the surface properties, at least for Hg. Bennett's (1970) result indicates that the surface plasma frequency may be shifted significantly from the classical value ω_p, depending on the surface charge distribution. Such inhomogenities may be responsible for the departures from Drude behavior found by Vanderhoff *et al.* (1971).

A sum rule has been applied to $\mathrm{Im}[\varepsilon^{-1}(\omega)]$, the energy loss function (e.l.f.). An alternate set of optical parameters may be obtained from the position and halfwidth of the e.l.f. peak. Drude theory places the peak at

$$\omega_l = [\omega_p^2 - (4\tau^2)^{-1}]^{\frac{1}{2}}, \tag{2.38}$$

and the halfwidth as τ^{-1}. Values of τ and ω_p, derived in this fashion and denoted τ^E, ω_p^E, are given in Table 2.4. Again there is general but not exact agreement. Note that the presence of the peak in the e.l.f. in the visible, and its influence upon the reflectivity, provides an immediate explanation of the golden color of these solutions without the necessity of invoking inter-band transitions (Thompson and Cronenwett 1967).

In the preceding paragraphs the optical data of Li–NH$_3$ solutions are examined from several viewpoints. In each case the data are consistent with the same simple theories which succeed in describing optical data obtained on more common liquid or solid metals. The differences between optical and low frequency determinations of the several parameters are no greater than those often found in pure metals and find ready explanation in terms of an effective mass m^* different from the free electron mass m and in terms of the background dielectric constant produced by the free NH$_3$ molecules. Surface effects also may be important.

It remains to be seen how the present results can be reconciled with the earlier data of Beckman and Pitzer (1961). They obtained reflectivities relative to Hg and attempted an interpretation in terms of bound electron-states as well as a free electron contribution.

Their expressions for ε_1 and ε_2 were thus of the form

$$\varepsilon_1 = \varepsilon_1^{\mathrm{D}} - \omega_{\mathrm{p}}^2 \sum_j \frac{f_j \tau_j (\omega^2 - \omega_{0j}^2)}{(\omega^2 - \omega_{0j}^2)\tau_j^2 + \omega^2}$$

$$\varepsilon_2 = \varepsilon_2^{\mathrm{D}} + \omega_{\mathrm{p}}^2 \sum_j \frac{f_j \omega \tau_j}{(\omega^2 - \omega_{0j}^2)^2 \tau_j^2 + \omega^2}$$

(2.39)

where $\varepsilon_1^{\mathrm{D}}$ and $\varepsilon_2^{\mathrm{D}}$ is the usual free electron term given by eqns 2.36 and 2.37 and the additional term represents the contribution from bound electrons. A difference between their data and the Drude theory arises only at concentrations below 8 m.p.m. and only in the energy range below 1 eV. We attribute this disagreement to the well-known difficulties with measurements of the optical properties of Hg (Bloch and Rice 1969) and dismiss the discrepancy as unimportant. It is, of course, possible that some of the problems that plagued Meyer and El Naby (1963), see Smith (1969), were present in any or all of the data discussed here. It is more profitable to turn from transport to magnetic and other properties.

2.8. Deductions from magnetic data

The usual treatment of nuclear magnetic relaxation data in metals is based on the Korringa relation, (Korringa 1950, Pines

1955) which in turn is based on the assumption that the nuclei are relaxed by degenerate, free electrons. In Na–NH$_3$ solutions this equation predicts a relaxation time for the Na nucleus near 1.0 s, about 15 times the observed value. Clearly another mechanism must operate. Duval *et al.* (1968) suggest quadrupolar interaction of the metal nucleus and the nitrogens. Similar relaxation times are found in concentrated sodium nitrate solutions. Rossini and Knight (1969) report quadrupolar relaxation to be relatively unimportant in pure liquid Na and Rb. Those ions, however, lack the solvation sheath. The increase of Knight shift temperature is similar to the increase of σ with temperature, and the two phenomena may well have the same explanation: the relaxation of the solvation sheath at higher temperatures (Lelieur 1972).

The proton relaxation times reported by Newmark, Stephenson, and Waugh (1967) are consistent with the less effective relaxation provided by free electrons in the case of the Na nuclei. They find relaxation times near 7 s, and attribute the relaxation to interactions with the free electron gas (via the Korringa relation). It is somewhat surprising to find agreement with the Korringa relation for the protons and not for the metal nuclei. An explanation appears possible in terms of the solvated ion model used for the conductivity. The nitrogens are close to the metal ion and thus can provide the quadrupole relaxation observed. Similar effects have been attributed to the solvation layers about Li$^+$ in aqueous LiCl solutions (Woessner, Snowden, and Ostroff 1968). At the same time the ammonias screen the metal nucleus from the free electrons and thereby reduce the free electron relaxation. The protons are on the outer edge of the solvated ion and may be relaxed by the free electron gas.

The increasing Knight shift found throughout the metallic concentration range (Pines 1955, McConnell and Holm 1957, Acrivos and Pitzer 1962, O'Reilly 1964, Lelieur and Rigny 1973) is consistent with metallic properties. The Knight shift $\Delta H/H$ at nucleus N_i is given by (Knight 1956)

$$(\Delta H/H)_{N_i} = 8\pi/3(\chi/N_0)\langle|\psi(N_i)|^2\rangle_{E_F} \qquad (2.40)$$

where H is the magnetic field, χ the atomic spin susceptibility, N_0 Avogadro's number, $\psi(N_i)$ the electron wavefunction at the nucleus N_i, and $\langle\ \rangle_{E_F}$ indicates an average over the Fermi surface at E_F. χ must be normalized to the atomic volume. By using the known values of χ, $\langle|\psi(N_i)|^2\rangle_{E_F}$ may be extracted from the data. Duval *et al.* (1968) report $\langle|\psi(N_1)|^2\rangle_{E_F}$ is proportional to the electron density at both the ^{23}Na and ^{14}N nuclei. The pseudo wavefunctions appropriate to the form factors used in the conductivity theory (Russakoff and Ashcroft 1970) (Schroeder and Thompson 1968) have not been applied to the calculation of $\langle|\psi(N_i)|^2\rangle$ as yet. The temperature dependence of $\Delta H/H$ is the same as that of χ, so that $\langle|\psi(N_i)|^2\rangle_E$ does not appear to change with T.

It is possible that the temperature dependences of both σ and χ arise from density-of-states effects as the values of the logarithmic derivatives of each are nearly the same ($\sim 0\cdot01$ eV) (Duval *et al.* 1968, Lelieur and Rigny 1973). We have adopted a different view in the model calculation of the conductivity. However, that model does not provide a ready value for $d\chi/dT$.

The asymmetric e.s.r. line (Catterall 1965; Chan, Austin, and Paez 1970) clearly shows the influence of delocalized electrons, as had already been mentioned. The asymmetry may be conveniently described in terms of the asymmetry parameter A/B, which is the ratio of positive to negative peak heights in the derivative representation of the e.s.r. line. If there is simply distortion of the r.f. field by eddy currents, then $1\cdot0 \leqslant A/B \leqslant 2\cdot7$. The lower limit occurs when the skin depth greatly exceeds the sample size, and the upper when very little penetration of the r.f. field occurs. When spin diffusion also occurs (Dyson 1955), the line shape becomes more complex and for a small skin depth A/B ranges between $2\cdot7$ and 19. Since the present data (Chan *et al.* 1970) lead to A/B $= 3\cdot0$ the line shapes probably are not affected by spin diffusion. Values of the spin-relaxation time T_2 are very close to those found in pure Na and must be attributed to spin–orbit coupling. The presence of the solvating sheath of NH_3 molecules is felt, however, since the T_2's of Na–NH_3 and Cs–NH_3 solutions are much alike. Though the ion core plays an important role, the solvent molecules modify the spin–orbit interaction so that spin–orbit effects do not increase from Na to Cs in solution as much as in the pure metals. Here again we have an example of the assertion concerning solute effects which introduced this chapter.

2.9. Comments on positron data

Though positrons provide a probe potentially sensitive to the electron environment, the high energies of the positrons make the technique less useful in practice. Furthermore, only very recently have analytical procedures independent of an assumed positron wavefunction been devised (Stroud and Ehrenreich 1968) and these procedures are yet to be applied to M–NH$_3$ solutions. Each of the groups which has examined the solutions has arrived at its own explanation and each explanation has its flaws. The difficulty is that the data, both as to particle lifetime and momentum distribution $N(k)$, are completely insensitive to concentration except below the 10^{-1} m.p.m. range. These observations are particularly difficult to reconcile with the assumed transition from extended to localized electron states below 8 m.p.m. (Chapter 4). In the present chapter only the more concentrated solutions will be discussed.

The $N(k)$ data even at 94 per cent Cs in NH$_3$ (Arias-Limonta and Varlashkin 1970) bear little or no resemblance to the free electron parabola usually found in liquid conductors (Kusmiss and Stewart 1967) Fig. 2.17. Varlashkin and Stewart (1966) attribute this discrepancy to the annihilation of positronium (Ps) from a bubble state, while McCormack and Millett (1966) assume the structure to be due to variations in the atomic portion of the electron wavefunction in an annihilation process which does not involve Ps. We examine each suggestion in turn.

Varlashkin and Stewart (1966) following Ferrell (1956, 1957) consider the Ps to be bound into a bubble by the repulsive interaction of Ps and NH$_3$. The bubble radius is fixed by balancing the zero point energy of the Ps against the surface tension of the solvent. The latter quantity is necessarily taken from macroscopic measurements though the application is to a bubble of radius 5 Å containing only a few molecules. The Ps wavefunction is then computed assuming the bubble to be a spherical well. The momentum distribution shows approximately the structure observed, including a hint of a high momentum peak near $k = 2 \cdot 5 \times 10^8$ cm^{-1}. The temperature dependence is somewhat less than would be expected for a simple bubble model and the change of surface tension with temperature.

McCormack and Millett (1966) make a much more complicated calculation. They assume the positron to be described by a constant wavefunction. The electron wavefunction then is derived from a potential reminiscent of that used in the conductivity calculation above. The positive ion is assumed to be surrounded by a spherical dipole layer. Because the dipole layer occupies that region of space where the atomic portion of the valence electron wavefunction is large the dipole layer is smeared to avoid introducing any discontinuities (and associated high-momentum wavefunction components) into the potential. At the same time Poisson's equation must be satisfied, and the effect of the ionic and valence electron fields upon the molecular polarizability included. After a self-consistent calculation the $N(k)$ shown in Fig. 2.26 was obtained. The assumed distance from the center of the ion to the outside of the solvent layer was $\sim 3 \cdot 5$ A.

The calculations yield much the same $N(k)$. Varlashkin's is difficult to reconcile with the unchanging $N(k)$ found upon freezing (a process which would seemingly collapse the bubble) and

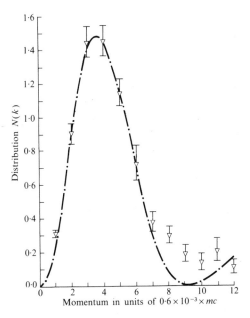

FIG. 2.26. Momentum distribution from positron annihilation in a 17 m.p.m. Li–NH₃ solution (triangles) and theory (line) (McCormack and Millett 1966).

with the large difference found between the solutions and pure NH_3. The pure solvent, after all, should provide electrons for Ps formation. The McCormack and Millett calculation first ignores a possible solvated state for the positron and second the disparity between extreme tight binding and high mobility. Yet their computation is much more within the spirit of the NFE calculation and of the known properties of the solid ammines (Chapter 7).

Majumdar and Warke (1967) have approached the problem of a lifetime calculation also from the Ps viewpoint. They attributed the quenching of the long lived component to ortho–para conversion of the Ps by electron–exchange collisions. No significance can be attached to their work because they ignored the fact that the metal atoms are ionized even at the lowest concentrations. Bhide and Majumdar (1969) consider the influence of solvated electrons, which do not exist at the concentrations of present interest. It is possible by combining lifetime and angular correlation data to determine the annihilation rate with valence (or free) electrons as a function of electron density or of r_s. Crowell, Anderson, and Ritchie (1966) determined this rate for direct annihilation with a free electron gas after making allowances for the influence of electron–electron interactions on the system dielectric constant. They find a minimum in the rate for r_s near 5, and a rapid rise for greater values of r_s. The minimum may not be real. There is competition, furthermore, with a possible bound (i.e. Ps) state. Fig. 2.27 shows rates determined for the solutions by Morales and Millett (1967) and rates for pure metals due to Bell and Jorgensen (1960) as well as the Crowell calculation and the Sommerfeld electron gas result. In view of the difficulty encountered in interpreting the angular correlation data, these irreconcilable results cannot be called surprising.

Cohen and Thompson (1968) have offered yet another suggestion, namely that the contribution of the positron wavefunction may be equal to that of the electron. As we shall see in Chapter III there is evidence that the positron (not positronium) state in dilute solutions is similar to that of the solvated electron. Solvation of the positron is also possible in the more concentrated solutions. A calculation similar to that made for Si and Al by Stroud and Ehrenreich (1968) is called for. Their calculation is based on a knowledge of the X-ray form factors of the solid and

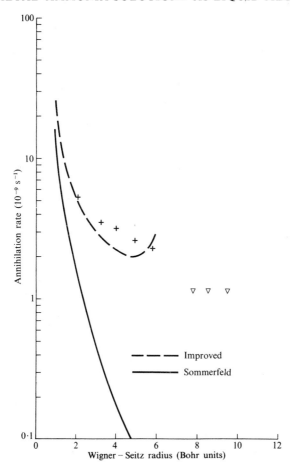

FIG. 2.27. Annihilation rates for positrons in several materials. The + represent, from left to right, Al, Li, Na, K, and Cs while the triangles represent three Li–NH₃ solutions (Morales and Millett 1967). The theory is due to Crowell, Anderson, and Ritchie (1966).

the relation of these form factors to the ionic potential felt by the positron. Such a calculation has not been made.

2.10. Other results and conclusions

Thermodynamic or mechanical data are not usually considered in NFE theory. Ashcroft and Russakoff (1970) 'used' density and compressibility data in developing the NFE theory of the conductivity. These data then are consistent with such a theory. Other properties, such as vapour pressure, viscosity, etc. show the trends to be expected for a metal (see Table 2.3) and also support the solvation model. As already noted the thermodynamic functions derived from the chemical potential show strong deviations from ideality. This is best illustrated by considering the mean square fluctuations in concentration (Bhatia and Thornton 1970) which are proportional to $\partial x/\partial \mu$ (Darken 1967). Ichikawa and Thompson (1973) have shown there are strong negative deviations from ideal behaviour when a compound or other cluster is present in the fluid. Fig. 2.28 shows data for several M–NH$_3$ solutions derived from vapour pressures (Marshall 1962). The positive deviations from ideal behavior at low concentrations are associated with phase separation (Chapter 5) while the dip near 15 m.p.m. for Na and Cs solutions, and the low value near 20 m.p.m. Li–NH$_3$ are correlated with solvation of the positive ion by a few (4–6) NH$_3$ molecules. These solvation numbers provide further confirmation of the model used in the conductivity calculation.

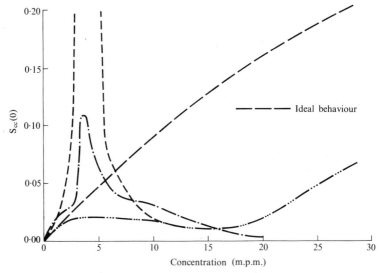

FIG. 2.28. The concentration correlation function $S_{cc}(0)$ for several M–NH$_3$ solutions (Ichikawa and Thompson 1973). The long dashed curve is the result expected for an ideal mixture.

Solutions of alkaline earth and rare earth metals in ammonia will be discussed in detail in Chapter 6. There are few differences between these solutions and the alkali metal solutions where comparisons are made at equivalent *valence electron* densities. There is every reason to believe that models similar to those presented for the alkalis will also be successfully applied to the solutions of divalent metals.

Solutions involving intermetallic compounds (Zintl phases) such as Li_4Pb will not be discussed (Zintl, Goubeau, and Dullenkopf 1931; Acrivos and Azebu 1971; and Schäfer, Eisenmann, and Müller 1973).

In conclusion, one finds that the properties of concentrated metal–ammonia solutions are qualitatively like those of simple liquid metals and may be quantitatively described in terms of a theory which describes the solutions as a metallic alloy of solvated metal and ammonia. In the next chapter more dilute, nonmetallic solutions will be described. The solvated ion will continue to be a useful construct and the 'solvated electron' will be adduced, as well.

REFERENCES

ABELES, F. (1966) (ed.). In *Optical properties and electronic structure of metals and alloys*, pp. 60; 175. North-Holland, Amsterdam.
—— (1972) (ed.). *Optical properties of solids*. North-Holland, Amsterdam.
ACRIVOS, J. V. and AZEBU, J. (1971). *J. Magn. Reson.* **4**, 1.
—— and PITZER, K. S. (1962). *J. phys. Chem.* **66**, 1693.
ADAMS, P. D., DAVIES, H. A., and EPSTEIN, S. G. (1967) (eds). *The properties of liquid metals*. Taylor and Francis, London.
ALLGAIER, R. S. (1969). *Phys. Rev.* **185**, 227.
—— (1970). *Phys. Rev. B* **2**, 2257.
ANIMALU, A. O. E. (1967). *Phys. Rev.* **163**, 557.
ARIAS-LIMONTA, J. A. and VARLASHKIN, P. G. (1970). *J. chem. Phys.* **52**, 581.
ARNOLD, E. and PATTERSON, A. (1964). *J. chem. Phys.* **41**, 3098.
ASCARELLI, P. (1968). *Phys. Rev.* **173**, 271.
ASHCROFT, N. W. (1966). *Phys. Letts* **23**, 48.
—— and LANGRETH, D. C. (1967a). *Phys. Rev.* **156**, 685.
——, —— (1967b). *Phys. Rev.* **159**, 500.
——, —— (1968). Errata: *Phys. Rev.* **166**, 934.
—— and LEKNER, J. (1966). *Phys. Rev.* **145**, 83.
—— and RUSSAKOFF, G. (1970). *Phys. Rev. A* **1**, 39.
BAILEY, K. E. and BOWEN, D. E. (1972). *J. chem. Phys.* **56**, 4809.
BALLENTINE, L. E. (1966). *Can. J. Phys.* **44**, 2533.
BECKMAN, T. A. and PITZER, K. S. (1961). *J. phys. Chem.* **65**, 1527.
BELL, R. E. and JORGENSEN, M. H. (1960). *Can. J. Phys.* **38**, 652.
BENNETT, A. J. (1970). *Phys. Rev. B* **1**, 203.
BHATIA, A. B. and THORNTON, D. E. (1970). *Phys. Rev. B* **2**, 3004.

BHIDE, M. G. and MAJUMDAR, C. K. (1969). *J. Phys. B* **2**, 966.
BILLAUD, G. (1972). Memoire, Lille. (Unpublished).
BLOCH, A. N. and RICE, S. A. (1969). *Phys. Rev.* **185**, 933.
BÖDDEKER, K. W. and VOGELSGESAND, R. (1971). *Ber. (Dtsch) Bunsenges. phys. Chem.* **75**, 638.
BOWEN, D. E. (1969). *J. chem. Phys.* **51**, 1115.
—— (1970). In *Metal–ammonia solutions* (eds J. J. Lagowski and M. J. Sienko), p. 355. Butterworths, London.
—— THOMPSON, J. C., and MILLETT, W. E. (1968). *Phys. Rev.* **168**, 114.
BRADLEY, C. C., FABER, T. E., WILSON, E. G., and ZIMAN, J. M. (1962). *Phil. Mag.* **7**, 865.
BRADY, G. W. and VARIMBI, J. (1964). *J. chem. Phys.* **40**, 2615.
BRIDGES, R., INGLE, A. J., and BOWEN, D. E. (1970). *J. chem. Phys.* **52**, 5106.
BUTCHER, P. N. (1951). *Proc. phys. Soc.* **64A**, 765.
CARUSO, J. A., TAKEMOTO, J. H., and LAGOWSKI, J. J. (1968). *Spectrosc. Lett.* **1**, 311.
CASTEL, J., LELIEUR, J.-P., and LEPOUTRE, G. (1971). *J. Phys. (Fr.).* **32**, 211.
CATTERALL, R. (1965). *J. chem. Phys.* **43**, 2262.
—— and MOTT, N. F. (1969). *Adv. Phys.* **18**, 665.
CHAN, S. I., AUSTIN, J. A., and PAEZ, O. A. (1970). In *Metal–ammonia solutions* (eds J. J. Lagowski and M. J. Sienko), p. 425. Butterworths, London.
COHEN, M. H., and THOMPSON, J. C. (1968). *Adv. Phys.* **17**, 857.
COOPER, B. R., EHRENREICH, H., and PHILIPP, H. R. (1965). *Phys. Rev. A* **138**, 494.
COULTER, L. V. and MONCHICK, L. (1951). *J. Am. chem. Soc.* **73**, 5867.
CROWELL, J., ANDERSON, V. E., and RITCHIE, R. H. (1966). *Phys. Rev.* **150**, 243.
CUSACK, N. E. (1963). *Rep. Prog. Phys.* **26**, 361.
DAMAY, P. (1973). In *Electrons in fluids* (eds. J. Jortner and N. R. Kestner), p. 195. Springer-Verlag, Heidelberg.
DAMAY, P., DEPOORTER, M., CHIEUX, P., and LEPOUTRE, G. (1970). In *Metal–ammonia solutions* (eds J. J. Lagowski and M. J. Sienko), p. 233. Butterworths, London.
DARKEN, L. S. (1967). *Trans. metall. Soc. A.I.M.E.* **239**, 80.
DEMORTIER, A., LOBRY, P., and LEPOUTRE, G. (1971). *J. Chim. phys.* **68**, 498.
DEVINE, R. A. B., and DUPREE, R. (1970). *Phil. Mag.* **21**, 787.
DEWALD, J. T. and LEPOUTRE, G. (1954). *J. Am. chem. Soc.* **76**, 3369.
—— —— (1956). *J. Am. chem. Soc.* **78**, 2956.
DRUDE, P. (1900). *Annln. Phys.* **1**, 566.
DUVAL, E., RIGNY, P., and LEPOUTRE, G. (1968). *Chem. Phys. Lett.* **2**, 237.
DYSON, F. J. (1955). *Phys. Rev.* **98**, 349.
EGELSTAFF, P. A. (1967). *An introduction to the liquid state.* Academic Press, London.
EHRENREICH, H. (1966). In *Optical properties of solids* (ed. J. Tauc), p. 106. Academic Press, London.
ENDERBY, J. E., NORTH, D. M., and EGELSTAFF, P. A. (1966). *Phil. Mag.* **14**, 961.
EVANS, R. C. (1964). *Introduction to crystal chemistry.* Cambridge University Press.
EVANS, R. (1970). *J. Phys., C (GB)* **3**, S137.
FABER, T. E. (1966). *Adv. Phys.* **15**, 547.
—— (1972). *Introduction to the theory of liquid metals.* Cambridge University Press.
—— and ZIMAN, J. M. (1965). *Phil. Mag.* **11**, 153.
FERRELL, R. A. (1956). *Rev. mod. Phys.* **28**, 308.

FERRELL, R. A. (1957). *Phys. Rev.* **108,** 167.
FREED, S. and SUGARMAN, N. (1943). *J. chem. Phys.* **11,** 354.
GARROWAY, A. N. and COTTS, R. M. (1973). *Phys. Rev. A* **7,** 635.
GLAUNSINGER, W. S., ZOLOTOV, S., and SIENKO, M. J. (1972). *J. chem. Phys.* **56,** 4756.
GRAPER, E. B. and NAIDITCH, S: (1969). *J. chem. Engng. Data* **14,** 417.
GUNN, S. R. and GREEN, L. G. (1962). *J. chem. Phys.* **36,** 368.
—— —— (1963). *J. Am. chem. Soc.* **85,** 358.
HARRISON, W. A. (1966) *Pseudopotentials in the theory of metals.* Benjamin, New York.
—— (1970). *Solid state theory.* McGraw-Hill, New York.
HEINE, V. and ABARENKOV, I. (1964). *Phil. Mag.* **9,** 451.
HODGINS, J. W. (1949). *Can. J. Res.* **B27,** 861.
HOGG, B. C., SUTHERLAND, T. H., PAUL, D. A. L., and HODGINS, J. W. (1956). *J. chem. Phys.* **25,** 1082.
HOLT, W. H., CHUANG, S. Y., COOPER, A. M., and HOGG, B. C. (1968). *J. chem. Phys.* **49,** 5147.
HOWE, R. A. and ENDERBY, J. E. (1967). *Phil. Mag.* **16,** 467.
HSUEH, L. and BENNION, D. N. (1971). *J. electrochem. Soc.* **118,** 1128.
HUSTER, E. (1938). *Annln Phys.* **33,** 477.
HUTCHINSON, C. A. and O'REILLY, D. E. (1970). *J. chem. Phys.* **52,** 4400.
—— and PASTOR, R. C. (1953). *J. chem. Phys.* **21,** 1959.
ICHIKAWA, K. and THOMPSON, J. C. (1973). *J. chem. Phys.* **59,** 1680.
JARZYNSKI, J., SMIRNOW, J. R., and DAVIS, C. M. (1969). *Phys. Rev.* **178,** 288.
JOANNIS, A. (1889). *C.r. hebd. seanc. Acad. Sci. Paris* **109,** 900; 965.
JOHNSON, W. C. and MEYER, A. W. (1932). *J. Am. chem. Soc.* **54,** 3621.
KIKUCHI, S. (1939). *J. Soc. chem. Ind. Japan* **42** (Suppl. binding) 15.
—— (1944). *J. Soc. chem. Ind. Japan* **47,** 488.
KNIGHT, W. D. (1956). In *Solid state physics* (Vol. 2) (eds F. Seitz and D. Turnbull), p. 93. Academic Press, New York.
KORRINGA, J. (1950). *Physica* **16,** 601.
KRAUS, C. A. (1907). *J. Am. chem. Soc.* **29,** 1557.
—— (1908). *J. Am. chem. Soc.* **30,** 653.
—— (1921*a*). *J. Am. chem. Soc.* **43,** 741.
—— (1921*b*). *J. Am. chem. Soc.* **43,** 749.
—— (1931). *J. Franklin Inst.* **212,** 537.
—— CARNEY, E. S., and JOHNSON, W. C. (1927). *J. Am. chem. Soc.* **49,** 2206.
—— and LUCASSE, W. W. (1921). *J. Am. chem. Soc.* **43,** 2529.
—— —— (1922). *J. Am. chem. Soc.* **44,** 1941.
—— —— (1923). *J. Am. chem. Soc.* **45,** 2551.
KUSMISS, J. H. and STEWART, A. T. (1967). *Adv. Phys.* **16,** 471.
KYSER, D. S. and THOMPSON, J. C. (1964). *J. chem. Phys.* **41,** 1162.
—— —— (1965). *J. chem. Phys.* **42,** 3910.
LEBOWITZ, J. I. (1964). *Phys. Rev.* **133,** A895.
LELIEUR, J.-P. (1972). Thesis, Orsay (Unpublished).
—— RIGNY, P. (1973). *J. chem. Phys.* **59,** 1142.
LEPOUTRE, G. and LELIEUR, J.-P. (1970). In *Metal–ammonia solutions* (eds J. J. Lagowski and M. J. Sienko), p. 247. Butterworths, London.
LINDHARD, J. (1954). *K. danske Vidensk. Selsk. Skr.* **28,** 8.
LO, R. E. (1966). *Z. anorg. Allg. Chem.* **344,** 230.
MACDONALD, D. K. C. (1962). *Thermoelectricity.* Wiley, New York.
MAJUMDAR, C. K. and BHIDE, M. G. (1968). *Phys. Rev.* **169,** 295.
—— WARKE, C. S. (1967). *Phys. Rev.* **162,** 247.

MAMMANO, N. and COULTER, L. V. (1969). *J. chem. Phys.* **50**, 393.
MARSHALL, P. R. (1962). *J. chem. Engng. Data* **7**, 399.
MAYBURY, R. H. and COULTER, L. V. (1951). *J. chem. Phys.* **19**, 1326.
MAYER, H. and EL NABY, M. H. (1963). *Z. Phys.* **174**, 280; 289.
MCCONNELL, H. M. and HOLM, C. H. (1957). *J. chem. Phys.* **26**, 1517.
MCCORMACK, K. and MILLETT, W. E. (1966). *Bull. Am. phys. Soc.* **11**, 201.
MCKINZIE, H. L. and TANNHAUSER, D. S. (1969). *J. appl. Phys.* **40**, 4954.
MEYER, A. and YOUNG, W. H. (1969). *Phys. Rev.* **184**, 1003.
MORALES, R. and MILLETT, W. E. (1967). *Bull. Am. phys. Soc.* **12**, 193.
MORGAN, J. A., SCHROEDER, R. L., and THOMPSON, J. C. (1965). *J. chem. Phys.* **43**, 4494.
MOTT, N. F. and JONES, H. (1936). *The theory of the properties of metals and alloys.* Clarendon Press, Oxford.
MUELLER, W. E. (1969). *Appl. Optics* **8**, 2083.
—— THOMPSON, J. C. (1969). *Phys. Rev. Letts.* **23**, 1037.
—— —— (1970). In *Metal–ammonia solutions* (eds J. J. Lagowski and M. J. Sienko), p. 293. Butterworths, London.
MYERS, W. R. (1952). *Rev. mod. Phys.* **24**, 15.
NAIDITCH, S., PAEZ, O. A., and THOMPSON, J. C. (1967). *J. chem. Engng. Data* **12**, 164.
NAIDITCH, S. and WREEDE, J. (1965). Final Technical Report, Contract NONR 3437(00). Unified Science Associates, Pasadena, California. (Unpublished, and private communication.)
NARTEN, A. H. (1968). *J. chem. Phys.* **49**, 1692.
NASBY, R. D. and THOMPSON, J. C. (1968). *J. chem. Phys.* **49**, 969.
—— —— (1970). *J. chem. Phys.* **53**, 109.
NEWMARK, R. A., STEPHENSON, J. C., and WAUGH, J. S. (1967). *J. chem. Phys.* **46**, 3514.
O'REILLY, D. E. (1964*a*). *J. chem. Phys.* **41**, 3729.
—— (1964*b*). *J. chem. Phys.* **41**, 3736.
PHILLIPS, J. C. and KLEINMAN, L. (1959). *Phys. Rev.* **116**, 880.
PINES, D. (1955). In *Solid state physics* (eds F. Seitz and D. Turnbull), Vol. 1, p. 367. Academic Press, New York.
—— (1962). *Many-body problem.* Benjamin, New York.
—— (1963). *Elementary excitations in solids.* Benjamin, New York.
PINGS, C. J. and WASER, J. (1968). *J. chem. Phys.* **48**, 3016.
POHLER, R. and THOMPSON, J. C. (1964). *J. chem. Phys.* **40**, 1449.
ROBINSON, J. E. (1967). *Phys. Rev.* **161**, 533.
—— and DOW, J. D. (1968). *Phys. Rev.* **171**, 815.
ROSSINI, F. A. and KNIGHT, W. D. (1969). *Phys. Rev.* **178**, 641.
RUSSELL, B. R. and WAHLIG, C. (1950). *Rev. scient. Instrum.* **21**, 1028.
SCHÄFER, H., EISENMANN, B., and MÜLLER, W. (1973). *Angew. Chem., Int. Ed. Engl.* **12**, 694.
SCHETTLER, P. D. and PATTERSON, A. (1970). In *Metal–ammonia solutions* (eds J. J. Lagowski and M. J. Sienko), p. 395. Butterworths, London.
SCHINDEWOLF, U. (1970). In *Metal–ammonia solutions* (eds J. J. Lagowski and M. J. Sienko), p. 199. Butterworths, London.
—— BÖDDEKER, K. W., and VOGELSGESANG, R. (1966). *Ber. (Dtsch) Bunsenges. phys. Chem.* **70**, 1161.
SCHMIDT, P. W. (1957). *J. chem. Phys.* **27**, 23.
SCHROEDER, R. L. and THOMPSON, J. C. (1968). *Bull. Am. phys. Soc.* **13**, 397.
—— —— (1969). *Phys. Rev.* **179**, 124.
—— —— and OERTEL, P. L. (1969)..*Phys. Rev.* **178**, 298.

SCHUMACHER, R. T. and SLICHTER, C. P. (1956). *Phys. Rev.* **101,** 58.
SMITH, N. V. (1969). *Phys. Rev.* **183,** 634.
—— (1970). *Phys. Rev. B* **2,** 2840.
SOMOANO, R. B. and THOMPSON, J. C. (1970). *Phys. Rev. A* **1,** 376.
STANEK, D. T. and LAGOWSKI, J. J. (1969). Private communication.
STERN, F. (1963). In *Solid state physics,* Vol. V (eds F. Seitz and D. Turnbull), p. 300. Academic Press, New York.
STROUD, D. and EHRENREICH, H. (1968). *Phys. Rev.* **171,** 399.
SUCHANNEK, R. G. (1966). *Rev. scient. Instrum.* **37,** 589.
—— NAIDITCH, S., and KLEJNOT, O. J. (1967). *J. appl. Phys.* **38,** 690.
TAKEUCHI, S. (1973) (ed.). *The properties of liquid metals.* Taylor and Francis, London.
THOMPSON, J. C. (1965). *Adv. Chem. Ser.* **50,** 96.
—— (1971). *Phys. Rev. A* **4,** 802.
—— (1973). In *Electrons in fluids* (eds J. Jortner and N. R. Kestner), p. 287. Springer-Verlag, Heidelberg.
—— and ALLGAIER, R. S. (1970). *Phys. Rev. A* **2,** 1103.
—— and CRONENWETT, W. T. (1967). *Adv. Phys.* **16,** 439.
—— and ORÉ-ORÉ, C. R. (1971). *J. chem. Phys.* **54,** 2279.
VANDERHOFF, J. A., LEMASTER, E. W., McKNIGHT, W. H., THOMPSON, J. C., and ANTONIEWICZ, P. R. (1971). *Phys. Rev. A* **4,** 427.
—— and THOMPSON, J. C. (1971). *J. chem. Phys.* **55,** 105.
VARLASHKIN, P. G. (1968). *J. chem. Phys.* **49,** 3088.
—— and STEWART, A. D. (1966). *Phys. Rev.* **148,** 459.
—— and THOMPSON, J. C. (1963). *J. chem. Phys.* **38,** 1974.
WEST, R. N. (1973). *Adv. Phys.* **22,** 263.
WHITTINGHAM, M. S. and HUGGINS, R. A. (1971). *J. chem. Phys.* **54,** 414.
WILSON, E. G. and RICE, S. A. (1966). *Phys. Rev.* **145,** 55.
WOESSNER, D. E., SNOWDEN, B. S., and OSTROFF, A. G. (1968). *J. chem. Phys.* **49,** 371.
WONG, W. S. (1966). Ph.D. Dissertation, University of California at Berkeley. (Unpublished.)
ZIMAN, J. M. (1960). *Electrons and phonons.* Clarendon Press, Oxford.
—— (1961). *Phil. Mag.* **6,** 1013.
—— (1967). *Phil. Mag.* **16,** 551.
—— (1973). In *The properties of liquid metals* (ed. S. Takeuchi), p. xiii. Taylor and Francis, London.
ZINTL, E., GOUBEAU, J., and DULLENKOPF, W. (1931). *Z. phys. Chem.* **154,** 37.

3

METAL–AMMONIA SOLUTIONS AS ELECTROLYTES

3.1. Introduction

WE begin this chapter with a reversal of viewpoint, considering solutions containing little metal and treating all the metal valence electrons as localized, though not localized upon the metal ion. Our language will be that of electrolyte solutions because the data will lead us to conclude that dilute solutions behave much as salt solutions do. The concentration range to be covered is from infinite dilution to $\sim 1\cdot 0$ m.p.m.

3.2. Electrical properties

As discussed in Chapter 1, there are many differences between water and ammonia as solvents. Nevertheless, most authors have used analogies with, say, NaCl solutions in water when discussing dilute Na solutions in NH_3. We will pursue this analogy in beginning the present chapter but will be obliged by the data eventually to abandon much of it. The increased conductivity observed upon dissolving metal provides the clue that ion and electron are separated. Thus we discuss conductivity data first, not only because more work has been done on this property than any other but because of the insight derived. As before we shall consider only alkali metal solutions. Also as in Chapter 2, we will take advantage of the fact that the properties of the solutions vary very little as the solute ion is changed.

3.2.1. *Electrical conductivity*

Much of the available data is a product of the extensive study of these solutions made by C. A. Kraus in the years 1900–1930. Fig. 3.1 shows some conductance data taken by Kraus (1921a,b), together with more recent results (Dewald and Roberts 1968). In these dilute solutions the equivalent conductance (Harned and Owen 1958)

$$\Lambda = (\sigma/N) \times 10^3, \tag{3.1}$$

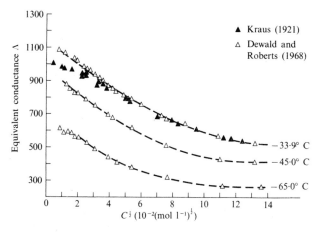

FIG. 3.1. Equivalent conductance Λ of Na–NH$_3$ solutions. See the caption to Fig. 2.1 (p. 15) for the notation in this and subsequent figures.

where N is the number of moles of metal per litre of solution, is more easily grasped. *If* all of the metal contributes to the conduction process then Λ is closely related to the mobility defined in Chapter 2. Fig. 3.2 shows Λ as a function of metal or salt concentration over a wider range of concentration (Berns 1964). The curves are seen to be qualitatively similar,

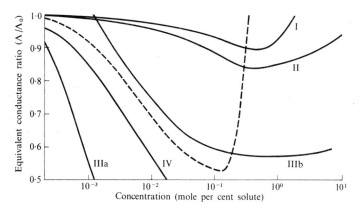

FIG. 3.2. Equivalent conductance ratio over the full dilute concentration range (Berns 1964). The solid lines refer to: I, HCl in H$_2$O; II, NaCl in H$_2$O; IIIa, KNH$_2$ in NH$_3$ (IIIB is $0.5 + \Delta/\Delta_0$); and IV NaCl in NH$_3$. The salt data are from Harned and Owen (1958) and Franklin (1910).

TABLE 3.1

Limiting ionic equivalent conductances[a]

Ion	Solvent	Temperature (K)	Λ_0 (Kohlrausch)
H⁺	H_2O	298	350
Li⁺	H_2O	298	38·7
Na⁺	H_2O	298	50·1
	NH_3	239	137
K⁺	H_2O	298	73·5
	NH_3	239	169
e⁻	NH_3	239	990[b]
OH⁻	H_2O	298	198
Cl⁻	H_2O	298	76·3
	NH_3	239	179

[a] References include Hnizda and Kraus (1949), Dewald and Roberts (1968), and Dewald (1969). [b] In Na–NH_3 solution.

each showing a broad minimum. Table 3.1 shows the various limiting conductances. Those for the metal–ammonia solutions are over three times greater than any observed with salts.

Onsager (1968) has remarked that the relatively shallower curve characteristic of the metal solutions may be indicative of important differences between metal and salt solutions (see below). Equivalent conductance data for salts also do *not* show so rapid a drop in H_2O solutions. The newer data of Dewald and Roberts (1968) weaken this point. Their data show a somewhat deeper minimum than the older, Kraus data. Both experimental groups used Pt electrodes and an a.c. bridge. Dewald and Roberts (1968) found Au electrodes gave the same results as Pt. The origin of these differences is difficult to establish as the data were taken 60 years apart. The most likely origin lies in cumulative decomposition errors in the Kraus experiments which employed successive dilution of a single solution. The modern workers used fresh solutions for each point as well as more stringent cleaning procedures. Presumably the conductances reported by Kraus were reduced by the amide formation in the reaction:

$$M^+ + e^- + NH_3 \rightarrow MNH_2 + \tfrac{1}{2}H_2 \qquad (3.2)$$

where M stands for the metal. Kraus was aware of the shortcomings of his technique and treated his low concentration results as lower limits to the actual Λ.

Kraus and Dewald and Roberts have investigated the effect of temperature T on the conductivity σ and Schindewolf, Böddeker, and Vogelsgesang (1966) have studied the effect of pressure P. It is found that σ increases with T, i.e. that $\gamma_T = \sigma^{-1} \, \mathrm{d}\sigma/\mathrm{d}T$ is positive; and σ decreases with P, i.e. that $\gamma_P = \sigma^{-1} \, \mathrm{d}\sigma/\mathrm{d}P$ is negative. The value of γ_T is a few per cent per degree and is only slightly dependent upon concentration. The pressure coefficient γ_P has the same concentration dependence as γ_T and is only a few per cent per bar. See Fig. 3.3, and also 2.5.

Potter, Shores, and Dye (1961) were unable to detect any enhancement of the conductivity when visible light illuminated a dilute solution contained in Pyrex, in contrast to an earlier report by Ogg (1946).

The observation of so large a conductance as in Fig. 3.1 requires us to assume free charges in the solution. The most likely source would be the ionization of the metal atom M. The positive ion (cation) would be the alkali metal ion M^+ and the

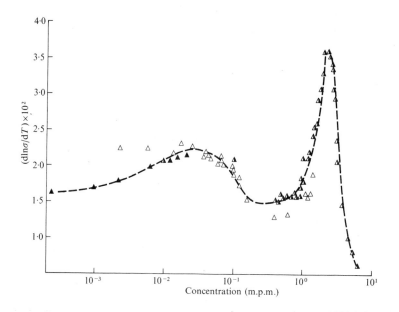

FIG. 3.3. The logarithmic derivative of conductivity σ with respect to temperature for Na–NH$_3$ solutions. The various symbols refer to various workers quoted by Dye (1970).

anion would be the metal valence electron e^- in a loose association with solvent molecules. In anticipation of the eventual conclusion e^- will be referred to as a 'solvated electron' or as an
'electron-in-a-cavity'. In subsequent discussion of the data, the
nature of the solvated electron will be discovered; for the moment, one may simply conclude that the process of solution
includes the following reaction

$$NH_3 + M \rightarrow M^+ + e^- + NH_3 \tag{3.3}$$

There is an immediate analogue in the reaction

$$NH_3 + NaCl \rightarrow Na^+ + Cl^- + NH_3 \tag{3.4}$$

which leads to the assertion that, in some sense, the metal
solutions may be considered as electrolyte solutions.

3.2.2. *Electrochemical measurements*

There have been a number of electrochemical investigations.
These include e.m.f. with transference (Kraus 1914), e.m.f. in a
galvanic cell (Russell and Sienko 1957; Ichikawa and Thompson
1973) transference number (Dye, Sankuer, and Smith 1960), and
chronopotentiometry (Gordon and Sundheim 1964). Except for
the galvanic cell work concerned with energetics, this work
primarily concerns carrier mobilities μ_i, which are related to the
ionic transference numbers t_i by

$$t_+ = \mu_+/(\mu_+ + \mu_-) \quad \text{and} \quad t_- = \mu_-/(\mu_+ + \mu_-).$$

Kraus's early work, which preceded his conductance measurements, led to the first estimates of the relative mobilities of the
solvated metal ion and solvated electron. He determined the
e.m.f. between solutions differing in concentration by a $2:1$ ratio.
Using the admittedly non-rigorous approach of an ideal solution
he determined the ratio t_-/t_+ to be near seven.

Dye *et al.* (1960) determined the metal ion transference
number by establishing an interface between the metal (Na)
solution and a salt (NaBr) solution, then measuring the speed at
which the interface moves when a current flows through the two
solutions (and through the interface.) The cation is the same in
both solutions, while the anion is the metal valence electron in
one case and the Br^- ion in the other. While they do not confirm
the concentration dependence of t_-/t_+ extracted by Kraus (who

assumed unity activity coefficients) the limiting ratio of seven was confirmed by Dye, Sankuer, and Smith (1960). More recent conductance data (Dewald and Roberts 1968) requires a slightly smaller ratio. Meaningful interpretation of transference data (Dye, Smith, and Sankuer 1960) led to the same association processes as does the interpretation of the minimum in Λ and will be deferred.

The anion (i.e. solvated electron) diffusion constant can be obtained by measuring the transition time for the depletion of excess solvated electrons at the cathode. This is the method of chronopotentiometric waves (Gordon and Sundheim 1964) and yields a value of $15 \times 10^{-5} \, \text{cm}^2 \, \text{s}^{-1}$ for the electron diffusion coefficient at infinite dilution. This compares well with the value computed from μ^- using the Nernst–Einstein equation.

There is no Hall effect data in the dilute range of concentrations (Nasby and Thompson 1970; Bellissent, Gerard, Longvialle, Meton, Pick, and Morand 1971).

We have already remarked (Chapters 1 and 2) on the effects of solution decomposition and in Chapter 2 the influence of 'non-ohmic' contacts on Hall effect studies was described. Schettler and Patterson (1970; Schettler, van Antwerp, Hamilton, Thilly, and Spear 1973) report extensive studies on electrode–solution effects similar to those reported by Nasby and Thompson (1968). The apparent non-linearities in electrode interactions are attributed to semiconducting behaviour of some surface layer on the electrode—probably amide. These layers persist in spite of rigorous cleaning procedures but can apparently be removed by prolonged electrolysis. This effect has often been missed by experimenters who do not closely monitor voltages; no account of its influence in, say, cell e.m.f. measurements is available. Dewald and Lepoutre (1954) do report the presence of an apparent thermal voltage in the absence of a temperature gradient; perhaps it is this effect.

3.2.3. Dielectric measurements

Microwave conductivities have been measured by Mahaffey and Jerde (1968). They placed cylindrical samples of the solutions along the axis of a microwave cavity. The shift in cavity resonance frequency and in cavity Q (loss factor) was analyzed to obtain the real and imaginary parts of the dielectric constants

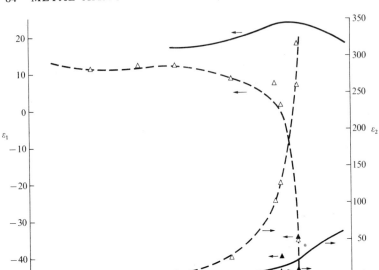

FIG. 3.4. Real ε_1 and imaginary ε_2 parts of the relative permittivity at 10 GHz open triangles (Mahaffey and Jerde 1968) closed triangles (Breitschwerdt and Radscheit 1971). The solid lines depict data on NaBr–NH$_3$ solutions due to Mahaffey and Jerde (1968).

shown in Fig. 3.4. The effect of added metal is opposite to that of added salt (Hasted, Ritson, and Collie 1948; Hasted and Tirmazi 1969). High frequency (215–70 GHz) permittivities have also been measured by Breitschwerdt (Breitschwerdt and Radscheit 1969, 1971, 1973; Breitschwerdt and Schmidt 1970; Apfel, Breitschwerdt, and Schmidt 1968; Breitschwerdt, Radscheit, and Wolz 1974) in Na–NH$_3$ solutions. These data supplement those of Hasted and Tirmazi (1969) but do not reveal the large structure in the spectra found by the latter authors. The frequency dependence shows the losses expected from the solvent together with a low frequency (10^{10} Hz) relaxation process. The analysis again involves association of M$^+$ and e$^-$.

3.2.4. Thermoelectric power and thermal conductivity

The only other electrical transport parameter studied in dilute solution is the thermoelectric power S studied by Dewald and Lepoutre (1954, 1956). Theirs is one of the most careful of all

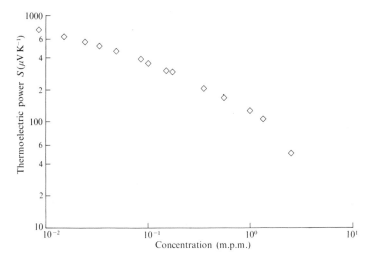

FIG. 3.5. Thermoelectric power S at 240 K (Dewald and Lepoutre 1954).

experiments in NH_3. They find a positive thermopower† S as large as 10^{-3} V K^{-1} and logarithmically increasing as a solution is diluted. The magnitude is essentially the same for Na and K solutions at 240 K, but is lower at lower temperatures for Na–NH_3 solutions (Fig. 3.5). There is a large and positive temperature dependence. If the Thomson coefficient is taken as the second derivative of e.m.f. with respect to temperature,† the results are as shown in Fig. 3.6, which is remarkably like the preceding figure. Dewald and Lepoutre (1956) also found that the addition of salt (NaCl) greatly increased S, as well as the concentration dependence of S.

A straightforward development of the irreversible thermodynamics of deGroot (1959) leads to a formal division of the thermopower into two terms, one arising from the heats of transport of the various carriers and the other arising from the concentration gradients (Dewald and Lepoutre 1956) (Ichikawa and Shimoji 1967). The latter is time dependent, as ionic

† The thermopower S is defined as

$$S(T) = [d\Phi/d(T' - T)]_{T' = T}$$

where Φ is the voltage measured, while T and T' are the temperatures of the two ends of the sample. S is positive if the hot electrode is positive with respect to the cold one.

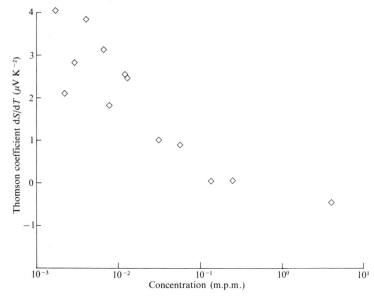

F IG. 3.6. The Thomson coefficient dS/dT at 240 K (Dewald and Lepoutre 1954).

mobilities are low enough that concentration changes cannot be effected in times less than those required to establish temperature gradients. Thus, a measure of initial and steady state Seebeck voltages permits both terms to be evaluated. The concentration term was found to be largest in dilute Bi–BiI$_3$ melts (Ichikawa and Shimoji 1969). Calculation of the thermopower requires knowledge of transference numbers, heats of transport, etc. The M–NH$_3$ data of Lepoutre is steady state data and leads to a heat of transport of -0.48 eV (Lepoutre and Demortier 1971). Both sign and magnitude are somewhat unusual.

Varlashkin and Thompson (1963) found that added metal caused an initial slight decrease in the thermal conductivity, followed by a large rise only when the solutions became metallic. There was little dependence on temperature or solute.

3.3. Magnetic properties

A wide variety of magnetic properties has been studied (for a review, see Catterall 1970). These include conventional static susceptibility measurements as well as magnetic resonance experiments: e.s.r. (Alger 1968), g-factor, electron relaxation times, n.m.r. on each available nucleus (metal, N, H), nuclear relaxation times, Knight shift, and the Overhauser effect. The consequences of substituting ND_3 for NH_3 have also been determined. The e.s.r. line is the narrowest known, at 20 mOe (Hutchison and Pastor 1953). The resonance experiments have been very potent tools for revealing the character of the solutions as they probe the microscopic environment of the spins without averaging over the whole system. The static data will be examined first, recalling that such data may not be clear-cut as their analysis is based on Wiedemann's rule (Myers 1952). This rule, as noted already in Chapter 2, is exceptionally naive, assuming no interaction between the magnetic moments. In view of the myriad of evidence for particle interactions in all kinds of solutions, and in view of the evidence for solvent modification by the solute ions, it seems unlikely to hold in any but the most dilute solutions. It is customary to abandon comparable assumptions (e.g. Raoult's law) as concentration increases but such sophistication has not appeared in the treatment of static magnetic susceptibility. It is fortunate indeed that independent determination of the spin susceptibility by e.s.r. techniques is available.

3.3.1. *Magnetic susceptibility*

In dilute solutions, static susceptibilities have been measured by Freed and Sugarman (1943) and by Huster (1938) using Gouy balance techniques. As noted earlier, the pure solvent is diamagnetic. On applying Wiedemann's rule one obtains the atomic susceptibility χ due to the added metal shown in Fig. 3.7. Freed and Sugarman (1943) used a modified Gouy method which yielded the susceptibility of metal in solution directly. Their analysis, however, still rests on the weak assumptions of the Wiedemann rule. The solid line in that figure shows the *spin-only* susceptibility expected for a set of non-interacting electrons with a number density equal to that of the metal valence electrons. It is seen that the two approach each other in the most dilute solutions at 240 K and that the data lie well below the curve at higher concentrations. Somehow the magnetic moments are being compensated. Nevertheless, the susceptibility in $K–NH_3$ solutions is systematically higher than that in $Na–NH_3$ solutions except at

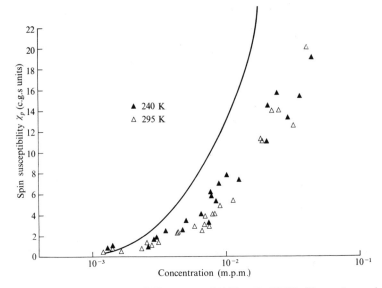

FIG. 3.7. The spin susceptibility χ_p at 240 K and 295 K (Demortier and Lepoutre 1969). The solid line is for free spins at 240 K.

the very lowest concentrations. The temperature dependence is, however, found experimentally to be exponential with higher susceptibilities at higher temperatures in both solutions.

3.3.2. E.s.r.

Hutchison and Pastor (1953*a,b*) reported extensive lowfield e.s.r. studies on K– and Na–NH₃ solutions in which *g*-factors, intensities, and susceptibilities, as well as line widths, shapes, and saturation were observed. Data could not be taken below 0·01 m.p.m. More recently, Demortier and Lepoutre (1969; Demortier 1970) have extended the data down to 0·001 m.p.m. The *g*-factor was found to be 2·0012±0·0002 independent of concentration and frequency and close to the *g*-factor of a free electron at 2·0023. The line has a Lorentzian shape. The area under the e.s.r. line yielded the values of the spin susceptibility shown in Fig. 3.7. In the e.s.r. experiments, diamagnetic contributions to the susceptibility are not involved. The general trends are much like those shown by the static susceptibility and lend credibility to the use of Wiedemann's rule there. Nevertheless, the r.f. susceptibility exceeds the static by a factor of almost

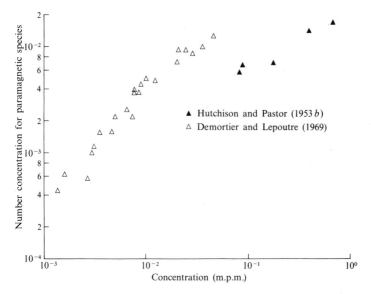

FIG. 3.8. Calculated concentration of paramagnetic species. All data are at 295 K.

two at concentrations near 0·1 m.p.m. (where each is more reliable). The susceptibility of K–NH$_3$ solutions is never greater than that of Na–NH$_3$ solutions, and becomes much less at room temperature; this trend opposes that seen in the static measurements.

O'Reilly (1964) has found indications that the spin susceptibility in Rb and Cs solutions exceeds that of solutions of the lighter alkalis. This appears to violate the rule that cations are unimportant. Again, the effects of spin compensation or of spin pairing (as in a singlet state) are observed.

The e.s.r. line is extremely sharp. Several experiments have been aimed at a determination of the origin of the narrowing. Pollak (1961) found that the relaxation times, which are inversely proportional to the line width, did not vary much with solute, though he observed the sequence $T_2(\text{Na}) > T_2(\text{K}) > T_2(\text{Cs})$ for the transverse relaxation time T_2 (Chan, Austin, and Paez 1970). O'Reilly (1969) found no solute dependence at 300 K below 0·1 m.p.m. except for Cs; indeed Li and K gave essentially identical results up to 2 m.p.m., see Fig. 3.9. The line-width ($\propto T_2^{-1}$) was exponentially dependent on the inverse temperature

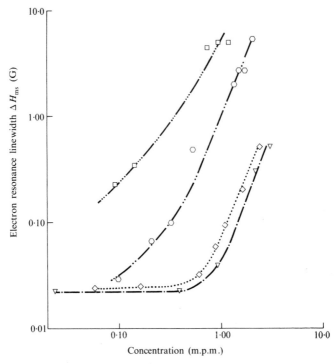

FIG. 3.9. The e.s.r. linewidth for several solutions at 300 K (O'Reilly 1969*b*).

T^{-1} (that is, Arrhenius behaviour) with an activation energy of 0·08 eV. The effect of substituting ND_3 for NH_3 was small (O'Reilly 1961). On the other hand, the substitution of ^{15}N for ^{14}N gave noticeably longer relaxation times (Pollak 1961). Pollak using a pulse technique found the spin lattice relaxation time $T_1 = T_2$ at all concentrations and temperatures and that the values were in the range 1–3 s. Typical data for K–NH_3 solutions are shown in Fig. 3.10. At low T the relaxation times increase, a maximum is then reached, and the more concentrated solutions show a decrease with T near room temperature. T_1 and T_2 are concentration independent except near the metal–nonmetal transition (Chapter 4). O'Reilly (1963), using a continuous wave (c.w.) technique, found $T_1 > T_2$. He attributed the difference to the influence of the hyperfine interaction between the electrons and the ^{14}N nuclei in the c.w. experiment, an influence which was

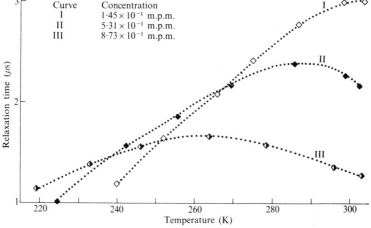

FIG. 3.10. E.s.r. relaxation times in three K–NH₃ solutions (Pollak 1961).

not present in Pollak's pulse experiment because of the relative magnitude of the times involved. On this assumption, O'Reilly derives a lower limit for the spin–lattice relaxation time of ^{14}N to be $(1\cdot2\pm0\cdot4)\times10^{-2}$ s which is somewhat larger than the directly determined (in pure NH₃) value of $(0\cdot4\pm0\cdot1)\times10^{-2}$ s. As the larger number is obtained by extrapolation the discrepancy is not unreasonable; nonhyperfine mechanisms appear at higher concentrations ($>0\cdot2$ m.p.m.).

3.3.3. N.m.r.

The n.m.r. lines of the various nuclei in the solutions are found to be shifted with respect to the same lines in (a) pure liquid ammonia for H and N; (b) the corresponding nitrates for Li and Na; and (c) the corresponding iodides for Rb and Cs. This effect has been called a Knight shift though that appellation is usually reserved for nuclei imbedded in a sea of free electrons. The Knight shift $K(N_i)$ is given by

$$K(N_i) = -(8\pi/3N_0)\,\chi_m|\psi(N_i)|^2 \qquad (3.5)$$

where N_0 is Avogadro's number, χ_m the spin susceptibility per nucleus N_i and $\psi(N_i)$ the electron wave function at the nucleus N_i in question, say, the proton. The combination of susceptibility and shift data will thus yield $\psi(N_i)$. Values of $\psi(N_i)$ will be

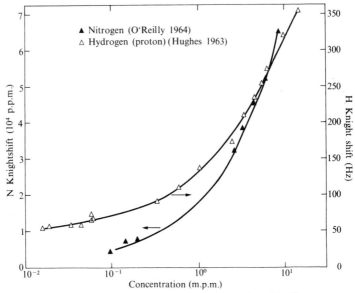

FIG. 3.11. Proton and nitrogen Knight shifts at 300 K.

presented for each nucleus when models are discussed. We examine the data for the nuclei of the solvent molecule first.

The Knight shifts at the hydrogen and nitrogen nuclei are shown in Fig. 3.11 measured relative to pure NH_3 (Hughes 1963; O'Reilly 1964). The effect of increasing temperature is to increase the Knight shift as shown in Fig. 3.12. The shift at the proton is very small, being only 1 p.p.m. at 1 m.p.m., and negative i.e. downfield. Carver and Slichter (1956) observed a positive Overhauser (1953) enhancement (an increase in the proton polarization when the e.s.r. line is saturated), which requires a contact interaction between unpaired electron and proton. Thus the small shift is apparently a Knight shift rather than a 'chemical' shift occasioned by, say, interrupted hydrogen bonding. This point is to be taken up again. Complications are found when $\psi(N_i)$ is determined, as the electron spin density turns out to be less than in pure liquid NH_3 according to Hughes (1963). Lambert (1968) confirmed this result with recent Overhauser effect measurements. A similar shift has been observed in $(ClO_4)_2[Cu^{2+}(NH_3)_6]$ (Wayland and Rice 1966). The similarity of hexaco-ordinated alkali and noble metal ions is surprising and

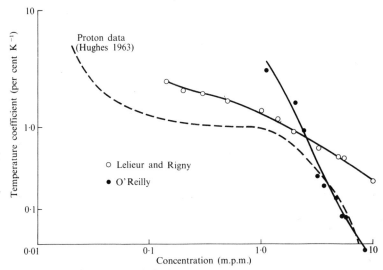

FIG. 3.12. Temperature dependence of Knight shifts of NH_3 atoms. Lelieur and Rigny (1973b) and O'Reilly (1964) do not agree (see §4.2).

points away from the solvated electron as an origin of the shift. Were it not for the Overhauser data this similarity would tend to discredit the Knight shift interpretation of the proton shift. It thus appears to be only coincidence that the proton is similarly affected by the two solutes.

The Knight shift at the nitrogen nuclei is two orders of magnitude greater than that at the protons and, as pointed out by Pitzer (1964), precisely proportional to the proton shift throughout the dilute range of concentrations. An immediate conclusion is that the same arrangement of electrons and solvent molecules is involved in each case. The temperature effect is also as before: the shift increases with increasing temperature. No change is produced when the metal is changed. We turn next to the various metal nuclei.

O'Reilly (1964a, b) has performed the most extensive set of experiments on Knight shifts at metal nuclei, though Acrivos and Pitzer (1962) and also McConnell and Holm (1957) have contributed. Some of O'Reilly's data are shown in Figs 3.13 and 3.14. The first figure shows the shift at nuclei of Li, Na, Rb, and Cs in ammonia at 300 K. The increase with atomic number is similar to that observed in the pure metals. Fig. 3.14 shows data obtained

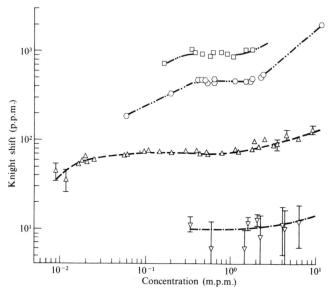

FIG. 3.13. Knight shifts at various metal nuclei at 300 K (O'Reilly 1964a).

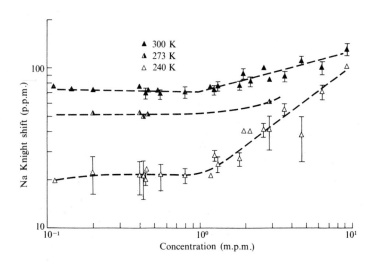

at three different temperatures for Na–NH₃ solutions. The Na–NH₃ data cover the widest concentration range at 300 K, and yet do not extend into the concentration range below the minimum in the equivalent conductance. There is some sign of a decrease in the shift at the lowest concentrations studied. Again the analysis will be based on eqn. (3.5).

Ample evidence for the hyperfine interaction between the electronic spin system and the various nuclear spins is provided, e.g. by the Knight shifts. Details of these interactions require investigation of relaxation times. Newmark, Stephenson, and Waugh (1967) have shown that coupling of proton spins to the electron assembly is an important source of proton relaxation. They determine the proton relaxation rate due to the dissolved metal by subtracting the relaxation rate for pure NH₃ from the measured relaxation rate of the solution. This difference is naturally small in the most dilute solutions, begins to rise only at concentrations near 10^{-3} m.p.m. (where spin pairing begins), and reaches a maximum near the concentration at which spin pairing is essentially complete. Unfortunately the correlation times extracted from the relaxation rates are so short (10^{-13} s) that distinction could not be drawn between the possible rotational and translational diffusion processes. Thus: (1) rotation of the solvated electron–NH₃ molecule complex; (2) exchange of co-sphere solvent with the bulk; and (3) electron tunnelling all may contribute. The latter process has been suggested as important by O'Reilly (1961).

The proton spin–lattice relaxation data in K–NH₃ of Newmark, Stephenson, and Waugh (1967) are among the scant data of any sort to extend to very low concentrations. They found relaxation in pure NH₃ to be inversely proportional to the viscosity as is expected. As metal was added the relaxation time decreased though the relation to the viscosity is maintained when the metal content is low. Values of order 5 s were obtained. Increasing temperature increased T_1. Itoh and Takeda (1963) have performed similar experiments, but do not agree with Newmark *et al.* As the Japanese report does not give an adequate description of experimental details the Newmark *et al.* work will be quoted in subsequent analyses.

The nitrogen relaxation time is known to be very short, comparable to that in pure NH₃ from the Overhauser work of

Lambert (1968) and also from the discrepancies in T_1 and T_2 between O'Reilly (1963) and Pollak (1961). Some linewidth data on the metal nuclei have been reported by O'Reilly, who finds T_1 in the range 0·06 to 0·2 ms in Rb and Cs solutions, but probably an order of magnitude longer in Na solutions.

High-resolution proton magnetic resonance (p.m.r.) measurements in K–NH₃ solutions have produced values of the ¹⁴N spin–lattice relaxation time, for those NH₃ molecules associated with the electron, of order 10^{-2} (Swift, Marks, and Sayre 1966; Pinkowitz and Swift 1971), that is, about two orders of magnitude below the proton relaxation time. This is consistent with the two order-of-magnitude difference in the Knight shifts and confirms that relaxation occurs through contact coupling (Solomon and Bloembergen 1956). Pinkowitz and Swift further conclude that the life time of a given NH₃ molecule in the neighbourhood of a solvated electron is about 2×10^{-12} s, a value quite close to both orientational relaxation times (Breitschwerdt and Radscheit 1969) and diffusion times (O'Reilly, 1969). That the number of NH₃ molecules involved at a given instant is large is indicated by the number of ¹⁴N with a contact shift. The number may be as large as 30 or 40. Neither temperature nor concentration dependence of the parameters could be resolved. Amide ion (eqn 3.2) was a major source of experimental difficulty due to rapid transfer of protons. The transfer effectively decouples proton and nitrogen to some extent and thus masks the ¹⁴N spin–lattice relaxation effect on the p.m.r. The addition of t-butanol (an acid) suppresses the NH_2^- concentration without interfering with the p.m.r. (Dewald and Tsina 1968; Birchall and Jolly 1965; Pinkowitz and Swift 1971).

Full discussion of the implications of the extensive resonance data will be deferred until other data have been presented. Note immediately the implication of spin pairing, of strong electron–nitrogen interaction, and of weak cation–electron interaction. Clearly the neighbourhood of a given 'solvated electron' does not include a metal ion.

3.4. Photons, etc.

The addition of metal to liquid ammonia produces three optical effects. One is a broad asymmetric absorption line which peaks near 0·9 eV. It has a high-energy tail which is responsible for the characteristic blue colour of the dilute solutions. The second effect is less well known. An absorption edge (or line; the high energy side has not been examined) in the near u.v. near 5 eV is enhanced and shifted to lower energies. The molecular vibration spectrum of the NH_3 molecule is also affected by the addition of metal.

3.4.1. Optical properties

Fig. 3.15 shows the shift of the ν_3 N–H bond stretching mode of the NH_3 molecule as observed by Rusch and Lagowski (1973) and by Beckman and Pitzer (1961). Here is a clear effect of the solute *upon* the solvent. It is somewhat surprising that all concentration effects are gone by 0·2 m.p.m. The data are complicated by the presence of other unresolved lines (ν_1 and $2\nu_4$) in the 3 μm (0·4 eV) portion of the spectrum where the ν_3 line is found and a complex computer program was used to effect resolution. The analysis indicates that both the symmetric and asymmetric N–H stretching fundamentals shift to lower energy with increasing metal concentration. The asymmetric bending overtone ($2\nu_4$) remains relatively unperturbed. The shifts observed by Rusch and Lagowski were the same for both Li and K; Beckman and Pitzer used only Na. No temperature effect was resolved in the 200–240 K range. A similar shift of ν_3 to lower energy was observed in highly concentrated salt–NH_3 solutions by Corset, Houng, and Lascombe (1968).

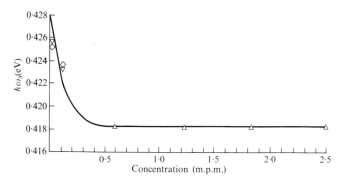

FIG. 3.15. Shift of ν_3 N–H bond-stretching modes of NH_3 in M–NH_3 solutions (Rusch and Lagowski 1973).

The effect of metal on the edge at 5 eV was apparently first observed by Burow and Lagowski (1965). These authors were engaged in a study of the effect of added metal on the 'charge-transfer-to-solvent' (CTTS) spectra of I^- in ammonia. The CTTS line was unaffected by the addition of metal but the edge was shifted (or, perhaps, the peak was enhanced) by the addition of metal as shown in Fig. 3.16 (this effect was pointed out to the author by W. H. Koehler). Clearly much more work must be done to characterize and identify this feature. Hart (1970) has pointed out that solvated electrons in H_2O seem to have a u.v. absorption similar to that found for F-centres in, say, KBr (Delbecq, Pringsheim, and Yuster 1951, 1952) and one may well expect such a phenomena in the NH_3 solutions. Note, however, that the influence of the I^- in the NH_3 data cited has not been assessed. No evidence is seen in the concentrated solutions (Chapter 2) which are nearly featureless out to 5 eV.

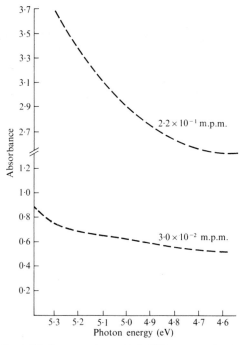

FIG. 3.16. Shift of CTTS spectra due to I^- in NH_3 as metal is added (Burow and Lagowski 1965).

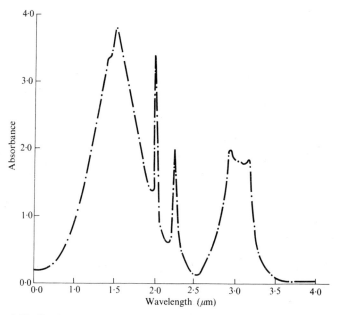

FIG. 3.17. Total absorbance of a dilute Na–NH₃ solution (Rusch and Lagowski 1973).

The blue colour of the dilute solutions and the absorption line responsible for it are very well known. Fig. 3.17 shows the absorption spectrum of a Na–NH₃ solution with both solvent lines and the 'solvated electron line'. Note the shoulder at 1.5μm which is due to the combination lines $2\nu_1$ or $2\nu_3$ (Rusch and Lagowski 1973; Burow and Lagowski 1965). In Fig. 3.18 is

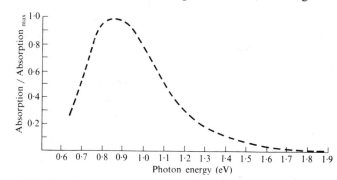

FIG. 3.18. Absorbance due to solvated electron only (Burow and Lagowski 1965).

shown the added absorbance due to the solute as a function of photon energy; the long tail running to 1·8 eV is the source of the colour. One of the most striking features of this line is that it is completely unaffected in so far as shape and location are concerned by systematically changing the solute from Li through Cs in the alkali metals, or from Li to Ca to Sr. The metals Eu and Yb also produce the 0·9 eV absorption when dissolved in ammonia. The same line is even found when tetraalkylammoniums are electrolyzed in liquid ammonia (Quinn and Lagowski 1968) or when electrons are injected by an accelerator Hart (1970). However, solutions wherein the anion is a simple ion, such as Cl^-, do not show any absorption in this portion of the spectrum other than that due to the solvent. Burow and Lagowski (1965) have emphasized that there is a weak pure solvent absorption line (combination line) at 0·9 eV which is the result of the superposition of low-lying vibrational absorption processes. As the line produced by the addition of metal is unchanged when ND_3 is substituted for NH_3, the major spectral feature is *not* to be assigned to the solvent.

Rentzepis, Jones, and Jortner (1973) have used a laser to follow in time the localization of an excess electron and find the optical spectrum to be fully developed in about 4×10^{-12} s.

Both the intensity and location of the absorption peak depend upon the amount if not the kind of metal: the intensity of the peak is proportional to the metal content (Beers' law); the coefficient of proportionality for divalent metals (e.g. Ca) is twice that for monovalent. The effect of metal concentration on the location of the peak (λ_{max} or $\omega_{max\ absorbance}$) and on the temperature derivative is shown in Fig. 3.19, as compiled by Koehler and Lagowski (1969) and by Rubinstein *et al.* (1973). The shift to lower energies occurs over a relatively narrow concentration range. For future reference it is important to note that this concentration range correlates with that in which the spin susceptibility decreases below the free spin value but lies above the concentration of the minimum in the equivalent conductance.

The line is also affected by temperature and pressure. The temperature coefficient $\delta_T = (\hbar\omega_{max})^{-1}\,\mathrm{d}(\hbar\omega_{max})/\mathrm{d}T$ is $-0\cdot17$ per cent deg^{-1} and again independent of solute. The pressure shift $\delta_P = (\hbar\omega_{max})^{-1}\,\mathrm{d}(\hbar\omega_{max})/\mathrm{d}P$ is about $+6 \times 10^{-3}$ per cent atm^{-1}. The proper treatment (Kestner, Jortner, and Gaathon 1973) of a

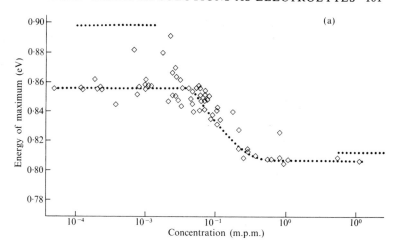

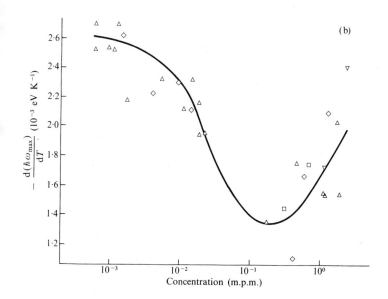

FIG. 3.19(a). Location of absorption maximum of solvated electron line at various concentrations, and due to various investigators (Koehler and Lagowski 1969). (b) Shift of temperature coefficient of absorption maximum with concentration. (Rubenstein, Tuttle, and Golden 1973).

temperature derivative requires consideration of both volume and (intrinsic) temperature effects (see also Chapter 4). The relations are

$$\frac{\partial(\hbar\omega_{max})}{\partial T}\bigg)_P = \frac{\partial(\hbar\omega_{max})}{\partial T}\bigg)_V + \frac{\partial(\hbar\omega_{max})}{\partial V}\bigg)_T \frac{\partial V}{\partial T}\bigg)_P.$$

The term on the left is what is measured at low concentrations (see Fig. 3.19) at the (low) vapour pressure: $-2{\cdot}0 \times 10^{-3}$ eV K^{-1}. The first term on the right is the intrinsic temperature term, while the second may be calculated from the pressure derivative as follows:

$$\left(\frac{\partial(\hbar\omega_{max})}{\partial V}\right)_T = \left(\frac{\partial(\hbar\omega_{max})}{\partial P}\right)_T \left(\frac{\partial P}{\partial V}\right)_T$$

$$= -\frac{1}{VK_T}\left(\frac{\partial(\hbar\omega_{max})}{\partial P}\right)_T,$$

where K_T is the isothermal compressibility. The thermal expansion coefficient is $V^{-1}(\partial V/\partial T)_P$. Combining these results yields

$$[\partial(\hbar\omega_{max})/\partial T]_V = -1{\cdot}5 \times 10^{-3} \text{ eV K}^{-1}$$

where the effect of thermal expansion amounts to $-5{\cdot}5 \times 10^{-4}$ eV K^{-1}. The temperature effect is thus partly a volume expansion effect. Note that the thermal expansion term above differs somewhat from the value reported by Kestner, Jortner, and Gaathon (1973) but rests on the sort of analysis used by Lelieur (1973) and described in more detail in Section 4.7.

The complete absence of a solute effect obliges us to either (1) assume that the optical transition occurs between electron states sufficiently remote from the metal ion as to be uninfluenced by the differences in pseudopotential and to depend only on the Coulomb field of the ion or (2) to be remote and independent altogether of the ionic potential. Case (1) might be a consequence of dielectric screening by the solvent or simply result from distance. It seems unlikely, however, that both monovalent (Na$^+$) and divalent (Ca^{2+}) ions might lead to the same spectrum. Case (2) is to be preferred at this stage, and is also consistent with conductance and magnetic resonance data.

Brief reports of the observation of a photoelectric effect have appeared. The threshold was found to be 1·5 eV in dilute solutions (Häsing 1940; Teal 1948). See also Schmeig (1939). There is considerable recent interest in photoemission and thermionic emission from metal solutions in a number of solvents (Baron, Chartier, Delahay, and Lugo 1969; Baron, Delahay, and Lugo 1970, 1971; Delahay 1971; Aulich, Baron, Delahay, and Lugo 1973). The photoemission work generally confirms the older papers but provides new information as well. The threshold seems nearer 1·8 eV than the 1·5 eV reported in earlier work. Fig. 3.20 shows the photoemission spectrum for a 1·2 m.p.m. K–NH₃ solution; comparable results are obtained at lower concentrations. The peak near 3 eV is attributed to the solvated electron (a bound-to-free transition) while the prominent peak at 4 eV is possibly associated with the absorption observed by Mueller and Thompson (1970) in reflectance measurements.

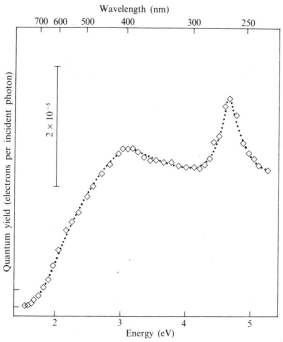

FIG. 3.20. Quantum yield of photoelectrons from a 1·2 m.p.m. K–NH₃ solution. (Aulich *et al.* 1973).

Amide effects have been ruled out in the photoemission. The 3 eV yield is proportional to the square root of the metal concentration below 0·2 m.p.m. (i.e. below the minimum in the equivalent conductance, Fig. 3.2). The yield increases less rapidly at higher concentrations. Both Na– and K–NH₃ solutions gave the same results. Backscattering and other effects based on the required transmission of the photoemitted electrons through the gas phase limits the accuracy of these results.

Thermionic emission has only been observed in M–HMPA solutions (see Chapter 8) (Baron, Delahay, and Lugo 1970) but such solutions are sufficiently similar to M–NH₃ solutions that comparable results might be expected. The work function in HMPA turns out to be close to 1·3 eV by both thermionic and photoelectric effects (Baron, Delahay, and Lugo 1971).

Three probes more energetic than optical photons have been directed at the solutions. One is X-rays, another is positrons, and the third electrons. We discuss these next.

3.4.2. X-rays

The early X-ray diffraction experiments of Schmidt (1957) showed the kind of structure expected for liquids and discussed in Chapter 2. The experiments were, unfortunately, carried out with solution surfaces exposed to air. There can be no doubt that decomposition occurred and any interpretation of the data must be questioned. Schmidt quotes similar data of Mechlin (1952) with similar limitations. Schmidt concludes that any scattering centres present contain metal ions and are ~15 Å in size. A more careful experiment has been reported by Brady and Varimbi (1964). They enclosed their samples in a polyethylene pillow. As with Schmidt the results were for small angles, and the scattering was found to be very much as for a normal liquid. Any inhomogeneities present must have had a size below about 6 Å.

3.4.3. Positron annihilation

Positron annihilation experiments have been done in dilute as well as concentrated solutions. The results obtained in dilute solutions for both angular correlation of annihilation gamma rays and mean life, are intermediate between those already reported in Chapter 2 and those observed in pure ammonia. Any concentration dependence has vanished by 10^{-1} m.p.m.

The lifetime data obtained by Morales and Millet (1967) contains the same 'glass correction' discussed before, and was shown in Fig. 2.27.

3.4.4. *Radiolysis*

When low-energy electrons are injected into liquid ammonia (radiolysis) the liquid turns blue. The optical absorption is found to be the same as that of a dilute metal solution. This observation clearly and unambiguously establishes the so-called solvated electron as the absorbing species and rules out any ionic influence.

In a typical pulse radiolysis experiment, the liquid to be irradiated is contained in a quartz cell with, say, a 0·5 mm window.

Electrons are accelerated to 3–4 MeV by a linear accelerator and injected into the liquid. Pulses are of near 1 μs length and deliver a dose of $10^8\,\mathrm{eV\,g^{-1}}$ to the sample. The transient blue colour is then analyzed with a spectrometer; the peak is identical to that of Fig. 3.18. In ammonia the colour has been observed to last up to many minutes (Dye, DeBacker, and Dorfman, 1970; Belloni and Fradin de la Renaudiere 1971). Solvated electrons form from the secondary electrons emitted by ionized ammonia. In the case of H_2O thermalization requires 10^{-13} s and hydration follows in 10^{-11} s (Bronskill, Wolff, and Hunt 1969). Comparable times are to be expected in NH_3. The following reactions occur:

$$NH_3 + radiation \rightarrow NH_3^+ + e^- \qquad (3.6a)$$

$$NH_3^+ + NH_3 \rightarrow NH_4^+ + NH_2^* \qquad (3.6b)$$

$$e^- + NH_3 \rightarrow e^- (solvated) \qquad (3.6c)$$

Loss of the solvated electron then proceeds by

$$e^-(s) + NH_4^+ \rightarrow H + NH_3 \qquad (3.7a)$$

$$e^-(s) + NH_2^* \rightarrow NH_2^- \qquad (3.7b)$$

$$e^-(s) + e^-(s) \rightarrow H_2 + 2NH_2^- (\text{very slow}) \qquad (3.7c)$$

$$e^-(s) + NH_3 \rightarrow H + NH_2^-. \qquad (3.7d)$$

The slow reaction of two solvated electrons in NH_3 is indicated by the stability of the M–NH_3 solutions; this is in strong contrast to hydrated electrons. Long lifetimes require a basic solution so as to suppress reaction (3.7d).

It is worth noting that hydrated electrons, i.e. solvated electrons in water, show many properties comparable to those of ammoniated electrons. Hydrated electron spectra have been observed in dense gaseous H_2O above the critical point (Michael, Hart, and Schmidt 1970). Similar experiments carried out in other solvents produce comparable though shifted spectra with a much shorter lifetime. These and other experiments involving solvents other than NH_3 will be taken up in Chapter 8.

3.5. Mechanical properties

A number of mechanical or thermodynamic properties have also been investigated in $M–NH_3$ solutions. These include: density, vapour pressure, viscosity, surface tension, adiabatic compressibility (sound speed), isothermal compressibility, and heat of solution. The perpetual conflict between quality and quantity of work makes many of the observations of limited value. In some cases, data were taken as a function of time, then extrapolated to zero time to obtain the 'true' value. In other cases, the concentration must be questioned, for reasons given in Chapter 1, though the property is accurately known. The properties are discussed in the order listed.

3.5.1. Density

The densities of metal–ammonia solutions are lower than that of liquid NH_3, except for Cs solutions. The volume exceeds the volume of the constituents in every case. The effect of added metal is opposite to that of added salt, as with the microwave relative permittivity and the adiabatic compressibility. It is customary to express the volume increase in terms of ΔV, the excess volume per mole of metal, where

$$\Delta V = [V_s - (V_1 + V_2)]/M. \tag{3.8}$$

In the equation, V_s is the solution volume, V_1 is the volume of the ammonia, and V_2 that of the solid metal; M is the number of moles of metal. Typical values of ΔV are shown in Fig. 3.21. The dotted line marks anomalously low values of ΔV reported by Evers and his co-workers (Brendley and Evers 1965), but disputed by Gunn (1967). In each case a dilatometer technique was used wherein the volume increase on the addition of metal was observed in a capillary. The experiments are sensitive to the usual

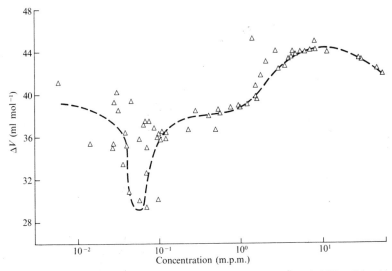

FIG. 3.21. Excess volume ΔV of Na–NH$_3$ solutions. Thompson (1967) collected these data from several authors. The dashed line is based on work quoted by Brendley and Evers (1965) but is *not* a 'best' fit.

problems of decomposition and temperature control. Gunn, in an elegant experiment, showed that dilution of a solution with an initial concentration above the minimum found by Evers did not produce a volume decrease and that dilution of a solution with an initial concentration at the minimum did not produce a volume increase. Thus he verified his earlier results and showed Evers *et al.* to be in error. Presumably, the error lay in the absence of an effort to account for the possible effect of bubbles produced in the warming and cooling of the sample during dissolution of the metal. Other data have been reported by Kraus and Lucasse (1921); Kraus, Carney, and Johnson (1927); Johnson and Meyer (1932); Kikuchi (1939); Hutchison and O'Reilly (1961); Demortier, Lobry, and Lepoutre (1971). Naiditch, Paez, and Thompson (1967) have covered a wide temperature range using a technique similar to that of Brendley and Evers (1965), though less precise. Then concentrations were assigned by comparing conductivities (simultaneously measured) to those of Kraus (1921). Fig. 3.22 shows some of Naiditch's results. The most striking point is that while ΔV is large at temperatures below the normal boiling point of NH$_3$ it tends toward zero above room temperature.

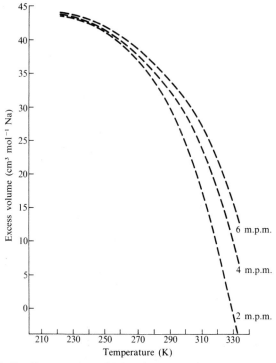

FIG. 3.22. Temperature dependence of excess volume (Naiditch, Paez, and Thompson 1967). Note that ΔV may well be negative above 320 K.

3.5.2. *Vapour pressure*

Vapour pressures were measured over a half-century ago by Kraus (1908), who thereby located some of the phase changes in the solutions; some of the solid compounds were found this way (see Chapter 7). The most recent, and presumably most precise (but, see Golden, Guttman, and Tuttle 1966) data are due to Marshall (1962). He studied solutions of Li, Na, K, Rb, and Cs at 238 K. The concentration range was from 0·03 to 50 m.p.m. The lowest concentration was reached only with Li and the highest only with Cs. Some data are shown in Fig. 3.23 as the ratio of the magnitude of the vapour pressure change ΔP to the vapour pressure P_0 of pure NH_3 at the same temperature; the quotient is then divided by the mole fraction. The actual quantity measured was ΔP. Three different differential manometers were employed,

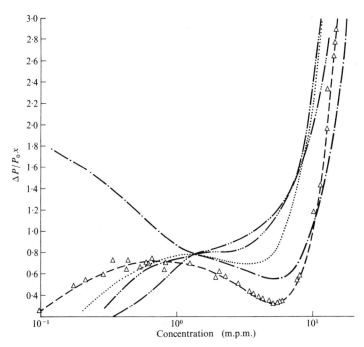

FIG. 3.23. Fractional pressure drop per unit concentration for several M–NH₃ solutions at 238 K; data are shown for only one (Marshall 1962).

depending upon the concentration range. Hydrogen evolved in the reaction of solute and solvent (eqn. 3.2) would be the most probable source of error. Marshall froze each solution, searched for residual H_2, and discarded any data obtained if H_2 was detected. Dewald's recent work indicates that some H_2 might well have been trapped in the frozen solution thus weakening our confidence in Marshall's work. In each case except Li, there is an initial slow rise followed by flattening, a slight dip and then a very rapid rise near 1 m.p.m. In the case of Li–NH₃, there is no sign of the initial rise and the general appearance of the curve is more like that of, say, a salt solution in liquid ammonia. Golden *et al.* reject Marshall's data out of hand because it differs so greatly from that typical of salts. As such differences occur so often in metal–ammonia solutions, we are inclined to view Marshall's data with no less or no more approval than any other.

3.5.3. *Viscosity and surface tension*

Viscosity measurements have been made by Kikuchi (1944), Nozaki and Shimoji (1969), Lobry (1969), and Hutchison and O'Reilly (1970).

As with the vapour pressures, the trends with concentration are opposite to those seen with dissolved salts: the viscosity of a metal–ammonia solution is less than that of pure ammonia while that of a salt in solution is greater. Kikuchi's data, at least, are very suspect since he extrapolated results which changed with time back to the time of mixing to obtain the alleged 'correct' value. His data, nevertheless, agree quite well with those of O'Reilly as may be seen in Fig. 3.24. Note that the concentrations are not particularly low.

Holly and Sienko (see Sienko 1964) have measured the surface tension of several metal–ammonia solutions and found it to exceed that of pure ammonia. The technique used was that of 'maximum bubble pressure' which involves the measurement of the pressure required to blow small bubbles in the liquid. At the point of maximum pressure a simple relation obtains between the pressure, bubble size (which is fixed by the tube used), density, and surface tension. Holly and Sienko found that their results

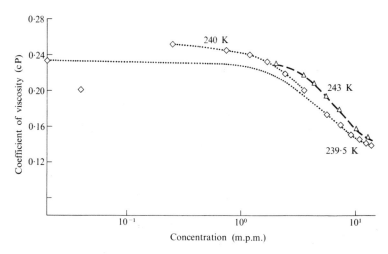

FIG. 3.24. Viscosity for several M–NH₃ solutions *near* 240 K (Lobry 1969 quoted by Lepoutre and Lelieur 1970).

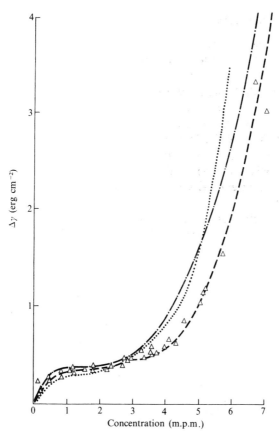

FIG. 3.25. Surface-tension increase at 233 K (Sienko 1964). The surface tension of pure NH_3 is $35·4$ erg cm^{-2} (Stairs and Sienko 1956).

could be reproduced only when the tube used to blow the bubbles was tipped with a noble metal. Though decomposition was believed to be the problem no effort was made to determine the precise cause. The results shown in Fig. 3.25 were obtained with a Pt tip. Only below $0·5$ m.p.m. do the metal solutions behave as do salt solutions. Above $0·5$ m.p.m. there is a flattening which is believed to be associated with the phase separation (Chapter 5) and not otherwise especially significant. As in nearly all systems the surface tension increases with T; the increase is insignificant except in the flattened portion of the curve.

3.5.4. *Compressibility*

Both adiabatic and isothermal compressibilities have been measured. The former was first determined by Maybury and Coulter (1951) and more recently by Bowen, Thompson, and Millett (1968), Bowen (1969), Parker and Bowen (1971) Bridges, Ingle, and Bowen (1970), Bowen (1970), Bailey and Bowen (1972), Franceware, Priesand, and Bowen (1972), and Thompson and Oré-Oré (1971). In each case the speed of a high frequency sound pulse was the quantity observed. Böddeker and Vogelsgesang (1971) have determined the isothermal compressibility as a part of other extensive pressure studies. We consider the sound speed first.

Exceedingly high precision is possible in acoustic time-of-flight measurements and accuracy is limited only by one's ability to determine the distance travelled by a pulse, as the elapsed time is easily measured on modern oscilloscopes. In practice, however, the precision has been limited by thermal fluctuations. Fig. 3.26 shows the apparatus used by Bowen (1969) for studies of K–NH₃ solutions. A quartz transducer generates a pulse of 10 MHz sound at A, the pulse travels down the quartz rod B and enters the solution at C. It is then reflected at D, returns to C, through B, and is finally detected at A by the same transducer. When the cell is empty, the sound is reflected at C because of the impedance mismatch between quartz and vacuum. Thus only the extra time to travel from C to D and back to C need be measured. The solution is made up in the side tube labelled E and the entire apparatus is immersed in a thermostat. The tapered end of the rod F prevents unwanted reflections. The adiabatic compressibility K_s is given by

$$K_s = C_s^{-2} \rho \qquad (3.9)$$

where C_s is the sound speed and ρ the density. Sound speed data are shown for all the metals in Fig. 2.19; compressibilities can be calculated when densities are available. The temperature coefficient of sound speed C_1 is shown in Fig. 3.27 for K–NH₃ solutions. The data in Fig. 3.28 are based on a smooth curve of density versus concentration which ignores the minimum reported by Brendley and Evers (1965). The compressibility deviates but little from that of pure NH₃ until 10^{-1} m.p.m., which

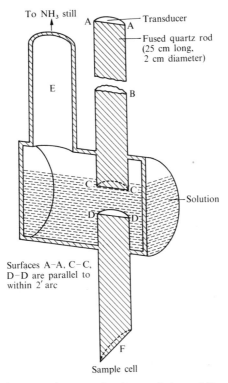

To NH₃ still

Transducer

Fused quartz rod
(25 cm long,
2 cm diameter)

Solution

Surfaces A–A, C–C,
D–D are parallel to
within 2′ arc

Sample cell

FIG. 3.26. Apparatus for measuring the speed of sound (Bowen 1969).

is well above the concentrations for irregularities in the conductance or susceptibility. The addition of metal increases the compressibility while the addition of a salt reduces the compressibility
(Garnsey, Boe, Mahaney, and Litovitz 1969). There are few
differences between the solutes, as usual in M–NH₃ solutions. See
also Andreae, Edmonds, and McKellar (1965) for a general
discussion.

The isothermal compressibility was obtained by measuring the
volume change as pressure was applied to a solution; data are
therefore only available above 100 atm and represent averages
over wide (100 atm) pressure ranges. Some of the data are shown
in Fig. 3.29 along with the ratio of adiabatic and isothermal
compressibilities γ. Boddeker and Vogelsgesang's isothermal data
for pure NH₃ lies about 10 per cent below that obtained by the

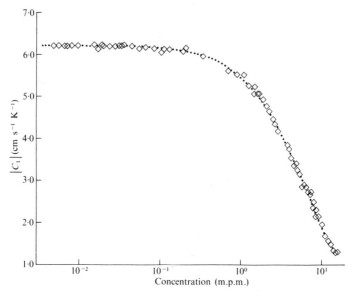

FIG. 3.27. Temperature coefficient of speed of sound (Bowen 1969).

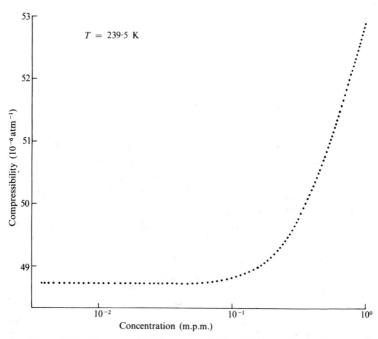

FIG. 3.28. Adiabatic compressibility K_s of K–NH$_3$ solutions. (Bowen, Parker, and Franceware 1973). Compare Fig. 2.2.

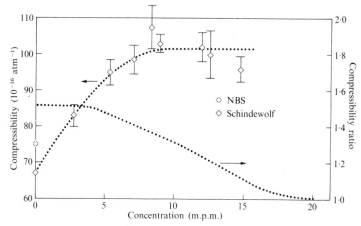

FIG. 3.29. Isothermal compressibility K_T (Boddeker and Vogelsgesang 1971) and ratio $\gamma = K_T/K_s = C_p/C_v$.

U.S. National Bureau of Standards Circular No. 142 (1923) at atmospheric pressure. One may safely assume that such a difference is due to the higher pressure at which they operated. The authors choose to represent the data as indicated by the line, though the errors bars would support a less smooth curve. The value of γ shown is based on values of κ_T adjusted from 100 to 1 atm. One thus obtains a smooth variation of γ from the 1·5 characteristic of many dielectrics to the 1·0 found in many liquid metals. This point will be pursued in Chapter 4.

3.5.5. *Heat of solution*

Gunn and Green (1962) have measured the so-called heat of solution; that is, the heat evolved (exothermic process) or absorbed (endothermic process) as the solute is dissolved by the solvent. They review the discrepancies found in older work, also reviewed by Coulter (1953). There are many possible sources of error, as with any calorimetric measurement. Here additional uncertainties are introduced by inadequate data on the densities and vapour pressures of the solutions as functions of concentration and temperature. Gunn and Green's (1962) data for Na–NH$_3$ solutions are shown in Fig. 3.30. The solution process becomes less endothermic as the concentration is increased and actually becomes weakly exothermic at higher temperatures.

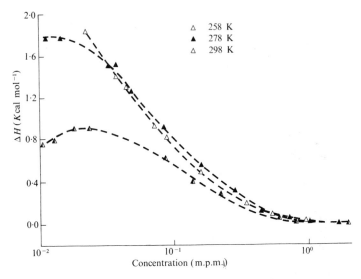

FIG. 3.30. Heat of solution (Gunn 1964).

There is a striking similarity between these results and the susceptibility data shown in Fig. 3.8; the similarity is not preserved at higher temperatures. It is tempting to ascribe the discrepancies to experimental error and we shall expand upon the consequences of the apparent correlation below. Table 3.2 contains 'average' heats of solution as compiled by Jolly; only Na is endothermic. The heats of reaction of the metals with ammonium

TABLE 3.2

Heats of solution of metals in NH$_3$ *at* 240 K[a]

Metal	ΔH (kcal/mole)
Li	−9·65
Na	+1·4
K	0
Rb	0
Cs	0
Ca	−19·7
Sr	−20·7
Ba	−10·0

[a] Jolly (1959).

salts are also given. If one visualizes the reaction as

$$e^- + NH_4^+ \rightarrow NH_3 + \tfrac{1}{2}H_2 \qquad (3.10)$$

then it is clear that the nature of the solvated electron is the same whatever the cation, as was already concluded from optical data.

3.6. Salt effects

The analogy between salt solutions and M–NH$_3$ solutions, with which we began this chapter, has led a number of investigators to add salts to M–NH$_3$ solutions. These mixed solutions have not yielded clarifying results in every case, due to problems with solubility limits, induced decomposition, and specific anion effects. (Studies with mixed metal solutes have had similar troubles.) It is not always clear whether the total metal (cation) content is the relevant parameter or whether the salt and elemental metal concentrations should be treated as independent variables so as to yield a three-dimensional graph of, say, conductivity as a function of Na concentration *and* NaCl concentration. The former is, at first, more appealing since Na$^+$ is the same whether derived from Na metal or NaCl. Yet the thrust of much of the data already presented is that the anion, particularly the solvated electron is dominant. One might thus try the Na and NaCl concentrations as variables. A sampling of results follows, necessarily without answers to all the above reservations.

In the range of considerable spin-pairing, adding NaCl increases the conductance, but not proportionally to the NaCl concentration (Berns, Lepoutre, Bockelman, and Patterson 1961). Somewhat different conductances were obtained with similar total Na$^+$ concentrations but different salt fractions. The addition of salt to a Na–NH$_3$ solution causes a large increase in the thermopower (Lepoutre and Dewald 1956) leading to values as great as $2\,\text{mV}\,\text{K}^{-1}$. The relation between thermopower and added salt is close to a square law in the salt concentration.

Catterall and Symons (1964) find that the e.s.r. spectra are considerably affected. The g-factor is shifted downward with added salt; the slope of the shift is approximately proportional to the ionic radius of the halide. As these data were taken in the range of considerable spin-pairing, considerable interaction exists among the solvated electrons and, thus, among other anions.

That is, these data indicate that solvated halide ions (at least I^-) enter into association equilibria interchangably with solvated electrons. This is confirmed by the relatively large Knight shift observed at the ^{127}I nucleus in K–KI–NH₃ solutions (O'Reilly 1969). Alternatively, one may conclude that associated species of the form minus:plus:minus do indeed exist and infer that solvated electrons may serve for either one anion (as in the present case) or both. Interestingly enough, the shift observed at the ^{127}I nucleus is approximately linear in the total K^+ concentration.

When the metal (and solvated electron) concentration is low, there is no optical effect from added salt (Quinn and Lagowski 1968) in the 300–2000 nm (0·6–6·0 eV) region in further confirmation of the assignment of the 1·5 μm absorption line to the solvated electron. Early reports of a shoulder are discounted (Clark, Harsfield, and Symons 1959; Catterall and Symons 1964; Gold and Jolly 1962). The solvent sheath surrounding each solvated species appears to retain its integrity. As noted earlier there are charge-transfer-to-solvent (CTTS) spectra in the u.v. attributable to the I^- ion.

3.7. Other systems

In the preceding pages of this chapter, a survey of the data available on the properties of solutions of alkali metals in liquid ammonia containing less than 1·0 m.p.m. metal has been presented. The amount of information is staggering and attests to the interest generated by these materials during the last century. As the data were presented it seemed reasonable to introduce two species (eqn. 3.3) as the primary constructs to be used in the analysis and to compare the metal solutions to solutions of salts. The properties of electrolyte solutions and the models used to describe them are next reviewed briefly. Again the literature is enormous and the concerned reader is referred to the standard texts and recent conference reports for a more careful survey. The present exposition is intended only to establish that the analogy exists and adduce only the broadest outline of the theory.

3.7.1. *Comparison with electrolyte solutions*

The salt solutions are discussed here because of the similarity between many of their properties (Robinson and Stokes 1959;

Conway and Barradas 1966) and those of M–NH₃ solutions; Fig. 3.2 is an example of such a case. Figs. 3.21, 3.24, and 3.28 show, contrastingly, properties which have trends opposite to those of salt solutions. In some cases, e.g. susceptibility, it is difficult to find analogues in simple solutions and one must dissolve salts containing, e.g. Mg in order to obtain magnetic solutions. Now the notion that similarities exist will be explored.

Salt solutions conduct and the equivalent conductance varies with concentration much as does that of the metal–ammonia solutions already shown in Fig. 3.2. The conductivity increases with temperature, inversely proportional to the viscosity. It is possible to separate out contributions to the conductance from individual ions, by comparing sequences of salts. Table 3.1 shows several limiting ionic (partial) conductances in water at infinite dilution $(x \rightarrow 0)$. Hydrogen and hydroxide ion have mobilities (or conductances) well above those for the other ions. There is a steady increase in mobility as one goes down the alkali or alkaline earth series. Surprisingly, the larger ions appear to move more easily. In ammonia, the alkali metal ions are somewhat more mobile than in water.

As the ions in salt solutions have only closed shells, there are no unpaired electron spins to produce a static paramagnetic susceptibility. There is, therefore, no data comparable to that indicating paired spins in the solutions. Unpaired spins can be obtained from free radicals, magnetic ions (e.g. Fe^{2+}), or other complex species, but do not provide the kind of analogy sought here. Though spin exchange has been observed in bimolecular collisions by Danner and Tuttle (1963), stable spin pairing does not seem to have been observed. N.m.r. data is available on solute and solvent nuclei. Relaxation times clearly indicate that the smaller ions, e.g. Li^+ bind the solvent molecules more tightly than do the larger ions. An immediate qualitative explanation of the decrease of mobility with decrease of ionic radius is thus provided: the smaller ion can move only by carrying some of the solvating molecules with it, thus rendering it effectively larger than the large (bare) ion which is only loosely bound to the solvent.

In those cases where solvating (bound) dipoles do not exchange rapidly with free solvent molecules and where the difference in the chemical shifts of the free and bound nuclei exceeds the exchange frequency it is possible to observe the differing environments of the two sorts of solvent and to determine, for example, the number of solvent dipoles bound to an ion. The solvation number depends upon the ion for a given solvent and upon the solvent for a given ion as well as on the observed property. For example the Mg^{2+} ion has solvation numbers of four, five, and six in H_2O, NH_3, and CH_3OH respectively. Swift and Lo (1967) infer that such changes would not occur in these solvents if cation–dipole forces dominated and that dipole–dipole effects or solvent–anion (i.e. ion pairing) interactions are present. They do not consider that the five-fold co-ordination in NH_3 could have simply been a mixture of four- and six-fold co-ordination numbers.

Wayland and Rice (1966) have studied chemical shifts in the co-ordination shells of Cu^{2+} and Ni^{2+} ions in liquid ammonia. Their results are of considerable possible significance to the understanding of metal–ammonia solutions. They believe that the solvated copper ion $Cu(NH_3)_6^{2+}$ may serve as a model for the solvated sodium ion. The protons in NH_3 coordinated to Cu^{2+} experience the same 'negative' Knight shift as has been observed in Na–NH_3 solutions. Furthermore, the coupling constants (which are proportional to the shifts) for the nitrogens and protons have a ratio of $-10\cdot6\pm1\cdot3$ in Na–NH_3 solutions and a ratio $-11\cdot7\pm1\cdot3$ in the ammonia molecules bound to the Cu^{2+}. The constancy of the ratio in each system suggests that there is only one source for both N and H shifts. The constancy of the shifts in the two systems (as the ratios are equal within experimental error) further suggests that the same mechanism is operating in each system. For the copper it is the co-ordination of the NH_3 to the Cu ion and the negligible hydrogenic character of the unpaired electron wave function in the co-ordinated ammonias that leads to the negative shift. The similarities between the two systems lead one to infer that the solvated sodium ion is the source of the shifts observed in Na–NH_3 solutions. The role of the sodium valence electron is then unclear. It seems more likely to this author that only a coincidence operates here.

Electrolyte solutions are generally colourless and the optical effects produced by the addition of, say, a salt to water are limited to specific transfer of charge from an anion to the solvent (e.g. I^- in NH_3). Such charge-transfer-to-solvent spectra (CTTS) are characterized by strong absorption lines. Other optical effects are sometimes seen in the IR. For example, the C–O stretching intensities are shifted in CH_3OH by the presence of Ca^{2+} or Zn^{2+}, though the sign of the shift is different for the two ions. In water solutions of salts such as NaCl, the intensity changes have been attributed solely to the anion. Frequency effects in the IR have been found to be small by Wall and Hornig (1967).

Before the full development of magnetic resonance techniques, studies of ion environments by measurements of microwave dielectric constants (as Mahaffey and Jerde 1968) were used. Hasted, Ritson, and Collie (1948) studied several aqueous solutions at $1\cdot25$, $3\cdot0$, and 10 cm. Solvent molecules bound to solute ions were unable to respond to the microwave field so that the dielectric constant was reduced relative to the pure NH_3 value (compare Fig. 3.4). Comparative studies of several salts led them to a set of hydration numbers not much different from those already considered. The loss term ε'' may be analysed using a Cole–Cole plot (Breitschwerdt and Schmidt 1970) to determine the effect of the salt on relaxation processes in NH_3. Relaxation times of order 10^{-11} s at 220 K are found and are consistent with viscosity limited processes. Solutions show two relaxation processes one of which is that of the pure solvent and the other involves ion pairs.

Positron annihilation experiments (Green and Lee 1964; West 1973) in salt solutions have often produced meaningless data due to the presence of dissolved O_2. In those cases where O_2 was unimportant two sorts of interaction between the Ps (i.e. $e^+ + e^-$) present have been observed. First, there is conversion of triplet Ps to singlet and thence a quenching of the Ps lifetime. The ions effective in this process are those having a free spin, such as Fe^{3+}, Cu^{2+}, Mn^{2+}, etc. The second possible interaction is a chemical reaction between e^+ and the solute. It is possible to regard the interaction between ion and Ps as providing electronic states in the 'Ore gap' and thus narrowing the energy range over which Ps is easily formed. The character of the solute is, of course, important.

3.7.2. Metal–molten salt solutions

Molten salts provide an extreme for electrolyte solutions, and the properties of metals dissolved in molten salts permit an analogy with metal–ammonia solutions (Sundheim 1964; Blander 1964). As an example consider Bi:BiI$_3$ solutions (Topol and Ransom 1963; Grantham and Yosim 1963; Ichikawa and Shimoji 1969). The electrical conductivity of the salt is increased by the addition of metal much as that of NH$_3$ solutions, except that the rapid upward break (see Chapter 2) occurs only at 50 m.p.m. The minimum in the equivalent conductance lies near 20 mole per cent Bi and is only 10 per cent of the value at infinite dilution (>4000 Kohlrausch). The temperature coefficient of conductivity is also much like that of the M–NH$_3$ solutions, particularly Cs–NH$_3$ (see Fig. 2.5), except that the maximum slope is at 40 m.p.m. and the sign change at 80 m.p.m. Thus the systems are much alike except when the high conductance of the solvent (BiI$_3$) is significant. Because of the high conductance of the BiI$_3$, more Bi is required ($\sim$50 m.p.m.) to produce a large increase in conductance.

The susceptibilities are diamagnetic throughout and decrease with increasing temperature. The temperature coefficient is greatest at those concentrations where an appreciable conductance change is first observed. The partial molar susceptibility of the Bi is weakly diamagnetic in dilute solutions and paramagnetic in concentrated solutions.

There is a net decrease in volume on mixing Bi with the salt because the added Bi occupies voids already present (Keneshea and Cubicciotti 1958).

It appears that the Bi valence electrons are localized at low concentrations; increasing concentration increases overlap and induces a metallic band above 70 m.p.m. There is evidence (Raleigh 1963) of hopping between mono- and tri-valent states of Bi at low concentrations.

3.7.3. *Comparison with excess electrons in non-polar fluids*

It should not be necessary to elaborate further similarities between electrolytes and metal solutions. One can gain more by a brief comparison to electrons injected into the liquefied noble gases He and Ar. There the odd electron exists in a bubble state (Springett, Jortner, and Cohen 1968) for He or in the conduction band for Ar (Jahnke, Meyer, and Rice 1971). The critical quantity is V_0: the ground state energy of an electron in the conduction band (measured relative to the vacuum). When V_0 is positive, energy is to be gained by a bubble; when V_0 is negative then the conduction band is the place for the electron to be. Mobility studies have been the primary tool for developing this model, as concentrations are too low to permit, say, optical or magnetic resonance measurements. The data in He are consistent with a cavity diameter of 20–30 Å with upwards of 100 He atoms involved in the wall. Even a classical mobility formula gives order-of-magnitude agreement with the $10^{-2}\,\mathrm{cm^2\,V^{-1}\,s^{-1}}$ mobility observed (Meyer, Davis, Rice, and Donnelly 1962; Eggarter and Cohen 1970). The basic reason for the existence of the bubble is an electron–atom repulsion derived from exchange forces. To this must be added surface tension, long-range polarization of the solvent, and the energy required to rearrange the fluid so as to allow for the bubble. One turns next to the models used in discussing salt solutions.

3.7.4. *Electrolytic solutions: fundamentals*

The constituents of electrolyte solutions are ions and solvent. In the most dilute solutions, at least, the ion–ion forces are unimportant so one first considers the interactions between solute ions and solvent molecules. The interaction depends upon the dipole moment of a solvent molecule. Solvent dipoles are oriented and bound by the Coulomb field of an ion. Adjacent dipoles, oriented by the same ion, repel each other and effectively weaken the ion–dipole force. Thus only a few layers, perhaps only a single layer, of solvent molecules will be bound to a single ion with binding energies exceeding thermal energies (Frank 1966; Ramanathan and Friedman 1971).

The ions together with the layer or layers of oriented solvent molecules are referred to as 'solvated ions'. Many experiments have been performed in attempts to determine what fraction of the solvent is bound into the solvation layers and what fraction, if any, is free. It emerges that the solvation layers in water and perhaps ammonia consist of 4–20 solvent molecules, packed more closely than in the pure liquid. The primary (i.e. first) solvation layer is oriented in the ion field. The second layer is often not bound at all and it is more accurate to speak of structural modification or solvent organization. Rapid exchange takes place between free and bound solvent, and many attempts to identify bound solvent molecules, say, by changes in their molecular

TABLE 3.3

Hydration numbers in strong electrolytes[a]

Ion	Solvation number
Li^+	3·0
Na^+	3·5
K^+	3·0
Rb^+	3·5
Cs^+	3·0
Ca^{2+}	6·0
Cl^-	1·0
Br^-	1·0
I^-	1·0

[a] Vogrin *et al.* (1971).

spectra have failed. This is rather surprising since the Coulomb field of the ion might be expected to distort the solvent molecule significantly. Exchange appears to weaken such effects. Nevertheless, n.m.r. spectra of aqueous solutions has led to experimental hydration numbers for many ions. Table 3.3 shows results recently reported by Vogrin, Knapp, Flint, Anton, Highberger, and Malinowski (1971). Note that anion solvation numbers are low, presumably as a consequence of steric hindrances. The close packing of the bound solvent molecules leads to an increase in the density of the solution over that of the pure solvent. That is, the volume of the solution is less than that of the solvent plus salt. If the solvated ions are treated simply as 'bare' rather than 'dressed' by the solvent, then one can attribute to them negative effective volumes. Conventional treatments of this sort are the foundation of the analysis of metal–ammonia data given for example by Gunn and Green (1962), and of the astonishment occasioned by the volume increase produced by the dissolution of metals in ammonia.

In strongly hydrogen-bonded liquids, such as H_2O, the orientation or organization process disrupts many hydrogen bonds and contributes thereby to reduced viscosity, etc. As noted, Hasted used measurements of the reduction of the microwave dielectric constant to estimate the number of solvent molecules hindered by charge–dipole forces from partaking in dielectric screening. The bound solvent is also less able to contribute to the vapour pressure and there is a vapour pressure lowering. Calculations of these properties require the assumption that the ions act independently. Such is not always the case.

Only at the very lowest concentrations do the added ions not interact; the solutions there are 'ideal'. Simple relations, such as Raoult's law, which gives the vapour pressure lowering as proportional to the solute fraction, apply. Isolated solvated cations and solvated anions are found only in extremely dilute solutions. As the electrolyte concentration is increased, Coulombic interactions among the ions become important. The standard treatment of the consequences of these effects begins with Debye–Hückel–Onsager theory. A simple Boltzmann factor is used self-consistently to determine the increase in the anion concentration in the neighbourhood of a cation (or vice-versa). This increase reduces the effectiveness of the Coulomb interaction by charge screening (as opposed to dipole screening). As a result the usual qr^{-1} interaction is replaced by $qr^{-1} \exp(-\lambda r)$, where λ is the screening constant (λ^{-1} is the screening length). In the linearized form of the theory λ is given for a 1–1 electrolyte by:

$$\lambda^2 = 4\pi n e^2 / kT, \tag{3.11}$$

where n is the ion number density. If the charge densities are such as to require the use of degenerate statistics then Thomas–Fermi screening is obtained. The Debye–Hückel theory leads to estimates of the free energy, etc. In the present context, an important result is that the equivalent conductance should go as the square root of the ionic concentration. We shall term the concentration range of such effects as that of 'simple association.'

The processes by which the equivalent conductance decreases are far from simple, however. If, as envisaged by Debye, Hückel, and Onsager, a cation has an excess of anions in its neighbourhood or 'atmosphere', then the cation will feel a drag when it attempts to respond to an external electric field (as also will the anions, of course). It seems clear that such a drag might produce the sort of decrease in Λ shown in Fig. 3.2 (Kay and Dye 1963). But it is also plausible that a cation–anion pair may result from these same associative forces. Each member of the pair presumably retains most, if not all, of its solvation layer. An ion may replace a solvating molecule (Vogrin et al. 1971; Ramanathan and Friedman 1971). The ion-pair, being neutral, will not respond to an external field. Further, pairing would seem to be more likely at higher concentrations. Again, a decrease in Λ would result. The ion-pair concept will be further examined below. Finally, it has also been suggested that the solution relative permittivity might increase with added salt so that any ion-pairs which might have formed at lower concentrations would dissociate. There are thus at least three processes which, singly or together, may produce a minimum in the equivalent conductance.

A brief development of the usual thermodynamics helps to quantify the concepts introduced above. Thermodynamically, the simplest approach is to consider the solute ions as constituting an ideal gas within the solvent. One may then write the Gibbs free energy G as:

$$G = N_1\mu_1 + N_2\mu_2 \tag{3.12}$$

where N_1, N_2 are the numbers of solvent and solute particles, and μ_1, μ_2 the respective chemical potentials. For the ideal case:

$$\begin{aligned} \mu_1 &= \mu_1^0 + RT \ln x_1 \\ \mu_2 &= \mu_2' + RT \ln x_2, \end{aligned} \tag{3.13}$$

where μ_1^0 applies to the pure solvent and μ_2' applies to the solute in the limit $x_2 \to 0$ rather than the pure solute. Experiment seldom reveals so simple a relation between $\mu = G/(N_1 + N_2)$ and $\ln x_1$ as eqn (3.13), so a correction term is added:

$$\mu_1 = \mu_1^0 + RT \ln x_1 + \mu_1^E, \tag{3.14}$$

where

$$\mu_1^E = RT \ln \gamma_1.$$

The excess chemical potential μ_1^E is the parameter giving the difference between ideal and observed behaviour; γ_1 is the activity coefficient. The *activity* a_i is defined by $a_1 = x_1 \gamma_1$ so that the final result becomes

$$\mu_1 = \mu_1^0 + RT \ln a_1, \tag{3.15}$$

which is much like (3.13). Such parameters as γ_i or a_i can be evaluated from the vapour pressure, for example. A square root dependence of a_2 on concentration is the variation expected from Debye–Hückel theory.

Ion pairs are said to exist when the activity fails to follow the square root law and as noted previously constitute a possible source of the minimum in the equivalent conductance. Two (unlike) ions are said to be paired if their separation is less than some radius R_i. The choice of R_i is to some extent arbitrary and the concept of an ion-pair is necessarily equally fuzzy. Though more sophisticated approaches (e.g. distribution functions) are clearly indicated, the ion-pair is a construct frequently adduced in physical chemistry. The solvation layers of the two ions may or may not be breached in the pairing process. When the bare ion is small (Li^+) the solvation layer is likely to preserve its integrity.

Ion pairing may be described by a pseudo-chemical reaction

$$A \rightleftharpoons B + C \tag{3.16}$$

and an equilibrium constant K assigned to the reaction. The equilibrium constant is defined as

$$K = \{(N_A - N_B)/N_C\}\gamma_\pm^2 \tag{3.17}$$

where N_i denotes the concentration of species i, and $\gamma_\pm$ is the average activity coefficient of the charged species; $\gamma_C \equiv 1 \cdot 0$ (Bjerrum 1926; Harned and Owen 1958; Robinson and Stokes 1959; Davidson 1962). K is proportional to the change in the chemical potential of the 'standard' state (See eqn 3.13) in the reaction. Statistical mechanical calculations are possible in simple cases, but in most cases K is extracted from experimental data on conductivity, susceptibility, etc.

H. S. Frank (1966) has pointed out that the Debye–Hückel (D.H.) theory is inadequate for more concentrated solutions. Difficulties arise with D.H. theory because screening lengths become too small to accommodate the charged particles which provide the screening and because the linearity imposed upon the Poisson–Boltzmann equation can no longer be justified (Robinson and Stokes 1959). The question is: what is the nature of the failure of the D.H. theory, for it is clear that the approximations of that theory cannot be valid above 10^{-2} molar. Corrections of all sorts have been attempted. Frank (1966) emphasizes that there is a concentration range in aqueous solutions where cube-root behaviour is observed for, say, activity coefficients. He considers, as an alternative to D.H. theory, that concentrated electrolyte solutions should be compared to molten salts where Coulomb interactions force the system into a structure with alternating positive and negative charges. One expects then cube root variations with concentration, rather than square root.

Stillinger and Lovett (1968) summarize the view of the phase diagram of an electrolyte as follows: in the most dilute region the structure appropriate to the Debye–Hückel theory obtains, with an excess of charge type B in the neighbourhood of type A. As the concentration is increased and electrostatic interactions become more important the structure tends toward one in which a given ion possesses concentric regions of both charge signs as in a molten salt. A phase separation may occur between the two different sorts of fluid if the interactions are sufficiently strong. Such separations have recently been reported for supercooled LiCl–H₂O solutions (Angell and Sare 1968). The discussion returns to NH₃ solutions.

3.8. Evidence of solvated electrons

The analogies with dissolved salts lead us to confirm our earlier conclusion that charged species exist in metal–ammonia solutions. The most likely source is, of course, the metal atom and it is reasonable to assume a dissociation of the form:

$$M + NH_3 \rightarrow M^+ + e^- + NH_3. \qquad (3.18)$$

The products of this reaction are a solvated metal ion, presumed identical to that obtained upon dissolving a salt, say NaCl, and an electron. The nature, or environment, of the metal valence electron once it leaves the atom is the subject of continuing controversy. We will first develop the currently most popular model for the electron in dilute metal–ammonia solutions, and mention only briefly those presently in disfavour.

Because of the relatively low ratio (7:1) between the conductances of anion and cation at infinite dilution (Kraus 1914), the electron cannot be free in the metallic sense discussed in Chapter 2. It compares more favourably with the anion in liquid He, than in liquid Ar. Because of the non-zero conductance, however, and because of the lack of variation of the optical absorption with solute, it seems likely that the electron is associated with one or more solvent molecules. The low density, or large excess volume, of the solutions led Kraus (1914) and later Ogg (1946a) to speak of the electron as residing in a bubble or cavity in the solvent.

Subsequent refinements were made by Lipscomb (1953) and by Kaplan and Kittel (1953), who also discussed the e.s.r. data in detail. Jortner (1959) has given the most extensive analysis of the electron-in-a-cavity and Section 3.23 follows his arguments and those of his co-workers (Jortner, Rice, and Wilson 1964; Jortner and Kestner 1970; Copeland, Kestner, and Jortner 1970; Kestner 1973). Before proceeding, however, one must attempt to distinguish between the notion of an electron-in-a-cavity, or solvated electron, and the properties of the specific anion described by Jortner. Simply by treating the electron as associated with the solvent one immediately gains a qualitative understanding of a wide range of phenomena:

(1) the low, but non-zero, conductance;
(2) the independence of the optical absorption upon solute;
(3) the presence of unpaired electron spins;
(4) the smallness of the solute Knight shift;
(5) the large size of the excess volume;
(6) the variation of viscosity with concentration;
(7) the correlation of H and N Knight shifts; and
(8) the relaxation of electron spins by ^{14}N interaction.

Items 1–4 result simply from the separation of ion and electron; items 1, 5, and 6 are connected with the presence of a cavity; and items 7 and 8 result from the presence of solvent molecules on the boundary of the cavity. Furthermore, the idea of a solvated electron follows very naturally from the common notion of solvated ions, with the requirement of a cavity or bubble the natural consequence of the low electron mass and concomitant high electron zero-point motion. One might almost regard the idea inevitable, considering the mind-set of the physical chemists involved in most metal–ammonia research. Recent discussions of solvated electrons in nonpolar media, such as He, have their roots in the NH_3 solutions but are conceptually much simpler as the interactions are dominated by electron–atom exchange repulsion (Jortner and Kestner 1970).

The most immediate use of the solvated electron is to provide an explanation for the volume excess. The ΔV defined by eqn (3.8) may now be resolved into contributions from cation and anion. The cationic contribution may be obtained from salt solutions, the residue is due to the solvated electron. As noted earlier, the effective volumes are conventionally negative. One is

TABLE 3.4

Summary of properties of model of solvated electron[a]

Property	NH_3 $V_0 = -0.5$, $N = 4$	H_2O $V_3 = 0.0$, $N = 4$
Ground-state energy (eV)	−2·40	−4·10
$\hbar\omega_{max}$ (eV)	0·94	2·6
$d(\hbar\omega_{max})\,dT$ (eV/K)	-0.5×10^{-3}	—
Band width at half-height (eV)	0·135	—
R_0 effective (Å)	3·0	0·0

[a] Copeland, Kestner, and Jortner (1970); Kestner and Jortner (1973); Kestner *et al.* (1973).

then faced with a large positive volume for the solvated electron, (Table 3.4). This volume corresponds to a spherical cavity of radius near 3 Å, independent of solute.

O'Reilly (1969) has recently used this large volume to explain his viscosity data, and perhaps to point the way to an explanation of other data, such as the compressibility, as well. His calculation is based on a treatment of the solution on a quasi-lattice model. Viscosity is then determined by the availability of vacancies in the lattice. Since salt ions fill vacancies, the viscosity increases upon the addition of a salt. The solvated electron, on the other hand, will cause an effective coalesence of several vacancies to form the electron cavity. Those molecules that surround the cavity will thus be able to move more easily as suggested by Samoilov (1957) and the viscosity will decrease to values below that of the solvent, as observed. The significant difference, here, between salt and metal solutions is merely the large size of the anion, that is, the presence of a cavity.

3.9. Association

Not only can one gain a number of qualitative explanations of the properties of M–NH₃ solutions from the concept of a solvated electron, but further rationalization of the data can be had by considering the association of the various charged species according to the reactions

$$e^- + M^+ \rightleftarrows M \qquad (3.19)$$

$$2M \rightleftarrows M_2 \qquad (3.20)$$

postulated before (Becker, Lindquist, and Adler 1956; Gold, Jolly, and Pitzer 1962). The nature of M and M_2 will be discussed in Sections 3.10.3 and 3.10.4. Detailed discussions of this process have been given by Arnold and Patterson (1964a) and by Golden, Guttman, and Tuttle (1965; 1966a, b). Unfortunately much of their numerical work is made void by the new conductivity data of Dewald and Roberts (1968) and the new susceptibility data of Demortier and Lepoutre (1969). Thus the conflicts listed below (Dye 1970) are in part the result of experimental difficulties.

3.9.1. Association in dilute metal–ammonia solutions

One begins by extracting the equivalent conductance of the solvated electron from conductance and transference number data. The ratio of equivalent conductances Λ^-/Λ^+ was found to be seven at infinite dilution. Dye finds the concentration variations shown in Fig. 3.31 and infers that ion pairing of the form (3.19) is required to interpret his data. The equilibrium constant deduced at 236 K was: $9 \cdot 2 \times 10^{-3}$. Evers could analyze Kraus' old data only if he assumed, in addition, that further association occurred:

$$M + M \rightarrow M_2$$

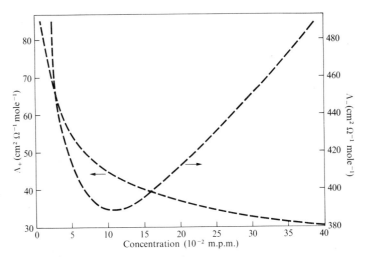

FIG. 3.31. Equivalent conductances of cation and anion (Dye, Sankuer, and Smith 1960).

with equilibrium constant K_2. His equilibrium constants were $7\cdot2\times10^{-3}$ and 27, for reactions (3.19) and (3.20) respectively. Dewald's more recent conductance data requires only reaction (3.19) and $K_1 = 3\cdot4\times10^{-3}$ at $-40\,°C$. Gunn's heat of solution data (Fig. 3.30) may also be used to check association constants. He obtained a fair fit using values of $K_1 = 8\cdot2\times10^{-3}$ and $K_2 = 18$ which were derived from Knight shift data. The association required by susceptibility will be considered next.

The spin data of Hutchison and Pastor (1953) show a distinct drop below the susceptibility expected for free non-interacting spins, and thus require species with paired spins such as M_2 or M^-. The equilibrium constants K_1 and K_2 derived from the 1953 magnetic data are near 30×10^{-3} and 98 respectively and stand in serious disagreement with the other data. Yet Thompson (1967) and others (e.g. Rubenstein, Tuttle, and Golden 1973) have pointed out a close correlation between the spin pairing and the shift in $\hbar\omega_{max}$ of the absorption curve with concentration, see Fig. 2.3. There is also a correlation with the heat of solution (Gunn and Green 1962; Tuttle, Guttman, and Golden 1966).

Demortier and Lepoutre (1969; Demortier, DeBacker, and Lepoutre 1972) have used their low concentration spin suscepti-bility data to extract equilibrium constants of 3×10^{-3} and 12 respectively, which clearly are much closer to the electrochemical data than those of Hutchison and Pastor. Indeed, they can fit both conductivity and susceptibility without considering M^-.

Arnold and Patterson (1964a) have pointed out a serious discrepancy. The data, they claim, cannot be made internally consistent using only eqn (3.19) and (3.20) and a third species $M^- = M + e^-$ is also required. Yet even their scheme has flaws; e.g. it fails to fit the heat of solution. Yet another association scheme is due to Golden, Guttman, and Tuttle (1965, 1966a, b), who are nevertheless unable to fit Marshall's vapour pressure data.

3.9.2. Dilemma in dilute metal–ammonia solutions?

This chaotic situation is not improved by the pressure studies of Schindewolf. He finds the equilibrium equation (3.20) not shifted with pressure, leading to the conclusion that the spin-pairing process is not volume sensitive.

Dye (1970) surveyed the various proposals at the Colloque

Weyl II. He reinforced the previous assertions that there were serious discrepancies in the values of the equilibrium constants as established by magnetic and electrochemical techniques. He further displayed inconsistencies which were model independent, thus compounding the dilemma. In spite of the long and detailed arguments which these discrepancies have engendered, it does not seem profitable to reproduce them at this time, as the newer data (especially Demortier and Lepoutre) promise to erase much of the controversy. The points of *agreement* are as follows. The isolated, solvated electron is essentially gone by 5×10^{-2} m.p.m., in favour of the paired electron species. The solvated electron–solvated metal ion ion-pair is present in very small concentration under all circumstances. For those occasions which require them, the equilibrium constants quoted by Demortier and Lepoutre will be used at 239 K:

Solute	K_1	K_2
Na	$2\cdot84\pm0\cdot03\times10^{-3}$	$12\cdot4\pm0\cdot8$
Cs	$3\cdot91\pm0\cdot03\ \ 10^{-3}$	$11\cdot4\pm0\cdot6$

Fig. 3.32 shows the concentrations of each species for Na–NH$_3$ at

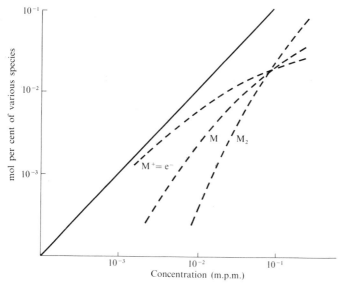

Fig. 3.32. Fractional concentration of various species in Na–NH$_3$ solutions at 239 K based on eqns (3.19) and (3.20). (Demortier, DeBacker, and Lepoutre 1972).

−34 °C. K_1 for K–NH$_3$ solutions must await better low concentration data. The problem, in part, is that little influence from the spin-pairing process is visible in the conductivity, though the presence of M_2 or M^- may not be ignored altogether in computing Λ. Yet the activity coefficients computed from e.m.f. (Kraus 1914) and vapour pressure (Marshall 1962) experiments do not agree with those expected from the equilibrium constants. The heat of solution (Gunn and Green 1962; Gunn 1964) changes with concentration in parallel with the susceptibility and optical absorption peak so requires the full equilibrium process (eqn 3.19, 3.20) for explanation. If there are no more errors such as those uncovered by Demortier and Lepoutre (1969) then it appears as if the definition of, say, ion-pair depends upon the method used to observe the pairing process.

Cohen and Thompson (1968) suggest it is the simple association *theory* which is at fault and that one must use the more sophisticated distribution function approach. They further suggest that the association process must go beyond even M_2 to very large complexes. Their analysis requires a better knowledge of the anion—the solvated electron—so we set association theory aside and return to the study of the nature of an isolated solvated electron.

3.10. Theory

The theory consists primarily of a description of the solvated electron, relatively little having been done on the other species (M, M^-, M_2). Indeed, there is strong controversy over the existence of a cavity containing two electrons.

3.10.1. *Basic theory of the solvated electron*

Fig. 3.33 is a schematic representation of a solvated electron. As with solvated ions, the charge–dipole forces would be expected to orient the solvent molecules and in the present case the oriented dipoles provide the potential which traps the electron. There is thus a close resemblance to the polaron introduced by Landau. Some have compared the solvated electron to an F-centre in an insulator.

As Copeland, Kestner, and Jortner (1970) (Kestner 1973) have recently emphasized the ground state energy contains contributions from both the electron and the rearrangement of the

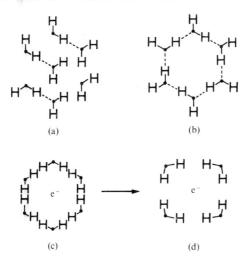

FIG. 3.33. Steps in the solvation of an electron in liquid ammonia (schematic). (Catterall and Mott 1969).

medium. The latter is positive for a localized electron, yet the total can be negative. Localization requires the free or quasi-free state to have a higher energy, V_0 than the localized state.

Fig. 3.33 was devised by Catterall and Mott (1969) to schematically illustrate the steps in solvation and permits the identification of the contributions to the medium energy. At a is the unperturbed medium with hydrogen bonds indicated by dashed lines and the nitrogen at the dot. At b a cavity formed in the fluid of a size near that of a single NH_3 molecule (Copeland, Kestner, and Jortner estimate a radius of $1·5$ Å). The presence of an electron at c tends to orient the dipoles in the first layer. Orientation is opposed by thermal agitation, by dipole–dipole repulsion, and by hydrogen–hydrogen repulsion (Bjerrum defects). The net result is the expansion of the first layer shown at d and a rotation of the NH_3 within the layer to reduce such effects. Long range polarization effects due to molecules outside the first layer can be treated on a continuum model (O'Reilly 1964; Land and O'Reilly 1967). In his early treatment of this problem, Jortner (1959) ignored the medium rearrangement energy. The results of the simple model are quoted here because of their transparency.

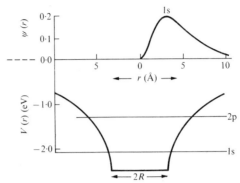

FIG. 3.34. Simplified model of a solvated electron well (Jortner 1959) together with the computed wavefunction and energy levels.

The solvent is assumed to be a continuous, homogenous medium with a low frequency relative permittivity ε_0 and a high frequency relative permittivity ε_∞. A spherical cavity of radius R_0 is excised from the medium and an electron placed within it. The electron interacts with both the permanent polarization of the medium and the electronic polarization. The potential produced by the former is given by

$$\varphi(r) = \frac{-\beta e^2}{r} \quad \text{for} \quad r > R_0$$

$$= \frac{-\beta e^2}{R_0} \quad \text{for} \quad r < R_0 \qquad (3.21)$$

where $\beta = [\varepsilon_\infty^{-1} - \varepsilon^{-1}]$. A large β means a deeper well and a stronger binding energy (see Fig. 3.34). The 1s and 2p wave functions are taken to be of the hydrogen-atom form:

$$\psi_{1s} = (\xi^3/\pi)^{\frac{1}{2}} \exp(-\xi r)$$
$$\psi_{2p} = (\zeta^5/\pi)^{\frac{1}{2}} r \cos\theta \exp(-\zeta r). \qquad (3.22)$$

The variational parameters ξ and ζ were evaluated by minimizing the energies of the ground (W_{1s}) and excited (W_{2p}) states:

$$W_{1s} = \frac{\hbar^2 \xi^2}{2m} - \frac{\beta e^2}{R_0} + \frac{\beta e^2}{R_0}(1 + \xi R_0)\exp(-2\xi R_0),$$

and

$$W_{2p} = \frac{\hbar^2 \zeta^2}{2m} - \frac{\beta e^2}{R_0} + \frac{\beta e^2}{R_0}(1 + \tfrac{3}{2}\zeta R_0 + \zeta^2 R_0^2 + \tfrac{1}{3}\zeta^3 R^3)\exp(-2\zeta R_0). \qquad (3.23)$$

The radii at which ψ_{1s} and ψ_{2p} are maximum, the Bohr radii, are ξ^{-1} and $2\zeta^{-1}$, respectively, on this model. Average radii r_0 are $\frac{3}{2}\xi$ and $\frac{5}{2}\zeta$, respectively.

The electronic polarization energy Π, is now introduced as a perturbation. Π will depend upon the state in question and is given by

$$\Pi = -\frac{1}{2} \int_{r_0}^{\infty} \left(\frac{eP_e}{r^2}\right) 4\pi r^2 \, dr, \qquad (3.24)$$

where r_0 is the average radius of the electron in the appropriate state and P_e is the electronic polarization given by

$$P_e = (e/4\pi r^2)(\varepsilon_\infty^{-1} - 1). \qquad (3.25)$$

Jortner thus obtains

$$\begin{aligned}
\Pi_{1s} &= -(e^2\xi/3)(1 - \varepsilon_\infty^{-1}) \\
\Pi_{2s} &= -(e^2\zeta/5)(1 - \varepsilon_\infty^{-1}).
\end{aligned} \qquad (3.26)$$

The total energy is $E = W + s$.

Several immediate comparisons can be made. The $1s \rightarrow 2p$ transition is presumed responsible for the blue colour characteristic of the dilute solutions; the energy is $0.8\,\text{eV}$. The energy derived from Jortner's model depends on R_0 as shown in Fig. 3.35. The closest agreement for NH_3 is obtained for $R_0 \simeq 3.0\,\text{Å}$. Cavities of radius 3 Å lead to an excess volume in close agreement with the volume attributed by Gunn and Green (1962) to the anion. A cavity of radius 3 Å leads to a ground state energy $1.6\,\text{eV}$ below the continuum, consistent with the heat of solution. Even the photoelectric threshold of $1.5\,\text{eV}$ fits the ground state level.

Other parameters come less freely: the shift of the absorption maximum with temperature and the shape and breadth of the line. Jortner (1959) originally suggested that the shift with temperature resulted from changes in the cavity size with temperature, as the changes in the dielectric constants were too small to produce the shift. However, it has already been shown in Section 3.9 that most of the observed temperature shift of $-1.5\,\text{eV K}^{-1}$ is due to thermal expansion of the fluid. The required shift in cavity radius dR_0/dT is only $3 \times 10^{-4}\,\text{Å K}^{-1}$ in this model. (But see Section 3.10.2 for more recent treatments.) The change in cavity

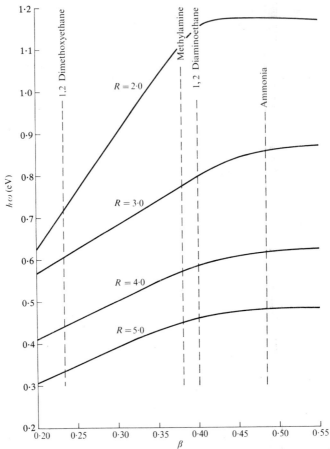

FIG. 3.35. Computed shifts of absorption maximum with relative permittivity, showing parameters for several solvents and several cavity radii in ångströms. The NH₃ data best fit is at $R = 3.0$ Å (Thompson 1967).

size is not easily compared with density data, as the effect of temperature on the sizes of the solvated metal ions is not well known. The best estimate of the change in the apparent volume is derived from Gunn's work which indicates $dR/dT = -2 \times 10^{-3}$ Å K^{-1} of opposite sign to the dR_0/dT above. Iguchi (1968) suggests that fluctuations in polarization produced by thermal agitation of the solvent molecules surrounding the cavity must also be included. He writes a Boltzmann factor to obtain the probability of a given dipole orientation in the Coulomb field of the electron. The gradient of the potential becomes weaker and the well deeper as a result. Iguchi chose *not* to apply his result to NH₃ solutions but to solutions in alcohols. One must infer that no great improvement was obtained.

Jortner (1959) suggested that the line shape was the result of transitions to levels higher than the 2p, e.g. 3p etc. Schindewolf, Catterall, and more recently Rusch, Koehler, and Lagowski (1970), have preferred to relate the line shape to fluctuations in cavity size. The latter authors have furthermore attempted to construct the line from the superposition of two or more *symmetric* bands. They were unable to find a unique set of bands which would reproduce the observed shape at all concentrations. Delahay (1971) suggests a combination of a Rydberg series with transitions to a band. See also Section 3.10.2.

Several authors have suggested that the line shape was the consequence of a variation in the radii of the cavities present. Rusch, Koehler, and Lagowski have made the most extensive analysis. They assume the band to be formed by the superpositon of *many* very narrow lines each the result of a 1s–2p transition in a cavity. However, the cavity sizes were assumed to differ. The relative intensities of the lines were taken from the transition probabilities appropriate to the cavity radius, and the frequency from eqn (3.22) and (3.26). The observed line shapes were then inverted to give the distributions of cavity radii shown in Fig. 3.36. Four significant features appear: (1) the most probable radius at infinite dilution is $3 \cdot 0$ Å; (2) the most probable radius shifts to larger R as x increases; (3) the distributions are symmetric in R; and (4) the width of the distribution is large. The energy required to polarize the ammonia is increased as the cavity size increases and is always much larger than thermal energies. The source of the fluctuations in cavity size is not easily found. Indeed the energies of cavity formation are not clear at all in this model.

Kajiwara, Funabashi, and Naleway (1972) have fitted the line shape and width to a model based on a finite range potential with only one bound state. The parameters of the model are fixed by matching the known locations of the absorption peaks. The shape and width are well reproduced. However, in view of the importance of long-range interactions in stabilizing the medium (Section 3.10.2) so as to allow the localized electron state to exist, this approach cannot be taken too seriously.

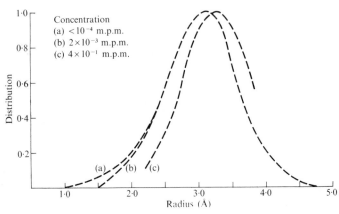

FIG. 3.36. Distribution of cavity radii required to produce shape and location of solvated electron absorption peak on Jortner's simple model (Rusch, Koehler, and Lagowski 1970).

3.10.2. *Further development of the theory of the solvated electron*

Attempts to improve upon Jortner's model have *not* been too successful until recently. Jortner, Rice, and Wilson (1964) attempted a self-consistent field calculation (S.C.F.) but were unable to fix the necessary parameters (cavity radius) completely. The absorption maximum for a $3 \cdot 2$ Å cavity was found to lie at $0 \cdot 9$ eV, above both experiment and simple model. O'Reilly (1964; 1971; Land and O'Reilly 1967) considered the first layer of solvent molecules explicitly and only treated the more remote solvent as a continuous medium. Their analysis, like Jortner's, led to absorption energies near those seen experimentally. O'Reilly, however, regards the cavity as a large bubble with 20 or more solvent molecules on its periphery. It would seem that the medium reorganization energy would, in such a case, be large and positive because of Bjerrum defects (Catterall and Mott 1969) so that the system of cavity plus solvent plus electron would be unstable. The Jortner model is therefore to be preferred.

O'Reilly (1964) was, however, able to show that the electron wave function could be made consistent with both proton and nitrogen Knight shifts. He found that a node in the wave function would lie near the protons Pitzer (1964). More recently O'Reilly (1971) has extended his treatment to the doubly occupied, spin-paired case.

In 1970 Copeland, Kestner, and Jortner made a major conceptual improvement. They considered not only the energy of the electron in the cavity but also the energy required to rearrange the medium so as to produce the cavity. They were thus able to give concrete answers to questions of stability, etc. Their approach has its foundations in the early Jortner (1959) model, in the calculations of O'Reilly (1964), and in the impressive work on excess electron states in nonpolar fluids (Springett, Jortner, and Cohen 1968; Jortner and Kestner 1970). The latter, particularly, focused attention on the stability of the localized electron state, the nature of electron–solvent interactions, and the modification (or organization) of the liquid structure in the neighbourhood of the solvated electron. Similar arguments were applied to a variety of solvents by Feng, Fueki, and Kevan (1973). Three parameters enter this model. The first is the free (or quasi-free) electron energy V_0; it is estimated to be near zero largely because NH_3 is isoelectronic with Ne (Springett, Jortner, and Cohen 1968). V_0 is <0 for the heavier rare gases and positive for He; for Ne it is presumed to be near zero. The second parameter is the size of the void R, which is taken as the distance from the centre to the hard-core surface of the NH_3 in the first co-ordination shell (see Fig. 3.37). The third is the co-ordination number z. These are interrelated since V_0 enters the total energy which is minimized

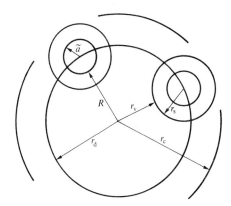

FIG. 3.37. Definition of the distances involved in the molecular model. r_v is the void radius of the cavity, r_s is the effective solvent radius, and $\tilde{a}$ is the effective hard core of the molecules located at a distance r_d from the centre of the cavity. The continuum begins at r_c (Copeland, Kestner, and Jortner 1970).

to obtain R. The energy of structural rearrangement first will be taken up.

The various contributions to the medium energy E_M have already been listed in connection with Fig. 3.33. There is the energy required to 'blow the bubble'. This includes work against the surface tension $E_{ST} = 4\pi\gamma R^2$, where γ is the surface tension, and the volume–pressure work $E_{PV} = \frac{4}{3}\pi R^3 P$. Secondly there is the long-range polarization of the medium as given by eqn 3.24 (actually the corrected version given by Land and O'Reilly (1967) (Kestner 1973) was used). Next there is the solvent–solvent interaction. Both dipole–dipole repulsions $E_{DD} = D\mu_{\tau}^3/r_d^3$ (where D is a constant, μ_{τ} the temperature-dependent, angle-averaged, induced-plus-permanent dipole moment, and r_d is as shown in Fig. 3.37) and proton–proton repulsions E_{HH} were included. Interactions at the quadrupolar level were neglected, as was the energy associated with hydrogen bonding. The response of the continuous medium beyond the first solvation layer to the charge distribution produces the polarization energy Π.

The electron sees a potential similar to that of Fig. 3.34 but distorted by the following effects. The entire potential scale is shifted by V_0, the energy of an electron free in the liquid as measured with respect to the vacuum state (Cohen and Thompson 1968). V_0 is not known and was taken $-0.5 < V_0 < 0.5$ by Copeland, Kestner, and Jortner. There is also the charge–dipole interaction which involves only that charge within the cavity and the effective dipole moment used in E_{DD}. These terms are different in the excited (2p) state from the ground state (1s). Thus the problem must be done self-consistently. Wavefunctions as in eqn (3.22) were used, and Π was added as before. The total potential seen by the electron was:

$$\begin{aligned}
\varphi(r) &= -z\mu_{\tau}e\langle\cos\theta\rangle/r_d^2 - \beta e^2/r_c, & 0 < r < R \\
&= -z\mu_{\tau}e\langle\cos\theta\rangle/r_d^2 - \beta e^2/r_c + V_0, & R < r < r_d \\
&= -(\beta e^2/r) + V_0, & r_d < r \qquad (3.27)
\end{aligned}$$

where V_0 acts in the entire region beyond the hard core radius, $\langle\cos\theta\rangle$ is given by a Langevin function, r_d is the distance to the centre of the NH_3 molecules of hard sphere radius a (see Fig. 3.37).

Total energies for $V_0 = 0$ are shown in Fig. 3.38; larger values

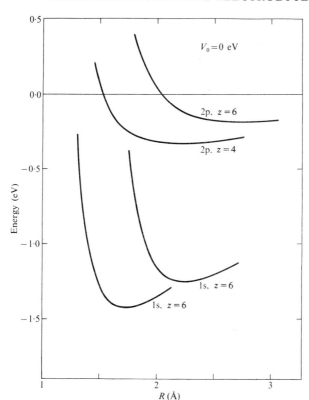

FIG. 3.38. The total energies of the ground (1s) and first excited (2p) states for the molecular model with $V_0 = 0$.

of z lead to similar curves at higher energies and larger values of R. Changing V_0 within the limits set (± 0.5 eV) does not make the ground state unstable; the system compensates for a $V_0 > 0$ by decreasing the values of ψ_{1s} outside the cavity. The differences in the depth of the minima between $z = 4$ and 6 (perhaps 8) are within the errors of the calculation. One may safely conclude that there is a *stable* cavity of radius $1.7 < R < 2.2$ Å.

The numerical results are somewhat disappointing when compared to the (fortuitous?) agreement found with the simple model (Jortner 1959). The absorption maximum is near 1.1 eV rather than 0.8 eV; the oscillator strength is $\frac{1}{3}$ that observed; the temperature dependence is $\frac{2}{3}$ that observed (Kestner, Jortner, and Gaathon 1973) and the heat of solution is 80 per cent of that found experimentally. The volume change however is in agreement in spite of the 1.7 Å radius. If the effect of the lower-than-average density in the first solvation layer is added to the cavity volume, an equivalent volume quite close to that based on a 3.2 Å spherical cavity is obtained. Cohan, Finkelstein, and Weissman (1974) have calculated the volume expansion with a slightly different treatment of the intermolecular interaction without altering the conclusions based on the Copeland, Kestner, and Jortner model.

Co-ordination numbers of four or six are consistent with Catterall's (1970) deductions and with the dielectric constants of Breitschwerdt and Radsheit (1971). The presence of nearby levels with $z = 6$ or 8 makes a distribution of cavity radii and co-ordination likely and may contribute to the high energy tail of the absorption line. One can also expect that asymmetric distortion of the cavity will not alter the total energy appreciably but may also lead to a high energy tail (Kestner 1973). These are thus at least three potential sources for the symmetry of the optical absorption lines; (1) a variety of cavity radii; (2) distortion of the cavity shape; and (3) transitions to states above 2p. Despite shortcomings the model appears to have answered most of the questions concerning the nature of an isolated, solvated electron.

3.10.3. *Remarks about the pairing*

The large size and diffuse nature of the solvated electron is important for association theory. The solvated electron is larger than, say, the solvated Li^+ ion. Thus in ion pairing or higher aggregates (e.g. M_2) the cation may fit into interstices between the electrons (Cohen and Thompson 1968). The co-ordination shells of the anion and cation probably contain common molecules. It is unlikely, however, that the cation replaces one of the co-ordinating NH_3 molecules. That there are only 4–8 solvent molecules involved with the electron is in apparent contradiction with the p.m.r. results (Pinkowitz and Swift 1971) but in general agreement with the treatment of solvated ions (Vogrin *et al.* 1971) shown in Table 3.3. One may infer that solvent molecules beyond the first shell feel the influence of the electron wavefunction, but the charge is sufficiently screened that their binding energy is near kT. The weak binding is certainly implied by the short $(10^{-12}\,s)$ lifetime of the NH_3 molecule in the sphere of influence (cosphere) of the electron.

Table 3.4 collects the parameters of the hydrated and ammoniated electrons.

While ion-pairing processes are readily comprehended within the Copeland, Kestner, and Jortner model, the problems of spin-pairing remain a cause of controversy. The problem is that some properties seem unaffected by spin-pairing, or only weakly affected, when naive treatments lead to contrary expectations. The parameters unchanged by spin pairing include:

(a) Volume. The equilibrium of eqn 3.20 is unaffected by pressure (Böddeker, Lang, and Schindewolf, 1970).

(b) Conductivity. The minimum is the equivalent conductance (Fig. 3.2) occurs when spin-pairing is essentially complete.

(c) The Knight shifts at the atoms of the NH_3 molecule. Pitzer (1964) notes that their ratio is also constant.

Other properties change somewhat, *as the spin-pairing proceeds*, and must thus be taken as consequences of the spin-pairing process. Some examples include:

(a) optical absorption peak (Fig. 3.19).

(b) heat of solution (Fig. 3.30).

Indirect evidence comes from e.s.r. and optical data on Eu–NH_3 solutions (Catterall and Symons 1965). No traces of Eu were found except as Eu^{2+} which contains an unpaired spin, yet there was broadening and asymmetry of the e.s.r. line as if a loose association of solvated electrons and cations existed. (That is, spin–spin interactions between Eu^{2+} and e^- were observed.) One may infer that any spin-paired species include a solvated ion for stabilization. The mixed metal–salt solution data indicate association processes involving unlike anions (e^- and I^-) where a shared set of solvating NH_3 molecules seems unlikely and thus supports the idea of solvent-separated electrons in the spin-pairing equilibria. Also the slight shift of the optical absorption peak and $d(\hbar\omega_{max}/dt)$ which accompanies pairing is more easily comprehended if the cavities remain separated.

One must ask: can two electrons occupy the same cavity? The *volume* of the spin-paired species is the volume of *two* unpaired electrons. One possibility, then, is that spin-pairing does not involve two electrons in the same cavity, (Cohen and Thompson 1968; Dye 1970). A number of theorists have discussed the two-electron cavity with varying degrees of sophistication† (Ogg 1946; Hill 1948; Land and O'Reilly 1967; Fueki 1969; Dexter and Fowler 1969; Fueki and Noda 1970; Kestner and Copeland 1970; O'Reilly 1971; Kestner 1973; Feng, Feuki, and Kevan 1973). The sum of this work is that it is likely that a spherical cavity containing two electrons would be stable against separation into two, isolated, solvated electrons, even when correlation is included. The two electron cavity in liquid He is in contrast unstable by about 0·5 eV, but may have a finite metastable lifetime (Dexter and Fowler 1969). The cavity is, however, much too small to fit volume data. The main difference between singly or doubly trapped electrons lies in the well depth (twice as deep for two as for one electron) and the influence of this depth on the medium rearrangement energy, particularly Π. Feng, Fueki, and Kevan (1973) predict a slight red shift on pairing as is observed (Fig. 3.19). A direct determination of the energy difference between singly and doubly occupied cavities is presently lacking, as is an explanation for the absence of a volume effect. The presence of one or two solvated cations in association with the spin-paired electrons (in one cavity or two) leads to various solute and salt effects. Rubinstein, Tuttle, and Golden (1973) believe M^- ($= M^+ + 2e^-$) to be the species containing the spin-paired electrons while the common assignment (Dye 1970) has been M_2, a neutral combination of two (separate) solvated electrons and two solvated ions. Much more work is clearly required, particularly with divalent solutes and with salts containing common-ions.

The doubly occupied cavity is a popular construct with those who attempt to explain the increasing conductivity above 0·1 m.p.m. seen in Fig. 3.1 (Onsager 1968; Cohen and Thompson 1968; Catterall and Mott 1969). Electron hopping from doubly to singly charged centres is an obvious and reasonable choice as the dominant mechanism. Judgment on this point will be withheld until the metal–nonmetal transition is fully explored in Chapter 4.

3.10.4. *The monomer*

Becker, Lindquist, and Alder (1956), and Das (1962) favoured a centro-symmetric version of the M species introduced in eqn

---

† It is ironic to note that Ogg proposed spin-paired electrons as an explanation for the superconductivity he *thought* he saw in solid Na–NH₃ solutions at 77 K (see Chapter 7) ten years before the BCS theory of superconductivity (Bardeen, Cooper, and Schrieffer 1957a, b) invoked the same construct.

(3.19). This species, called the *monomer* is envisioned as a solvated metal ion with an electron orbiting on the protons of the molecules in the first solvation layer. Though many of the known properties of M–NH₃ solutions can find adequate explanations in terms of this neutral species, it fails to provide a way to understand the increase in conductivity produced by the addition of metal to NH₃. The possibility that it exists as the neutral product of eqn (3.19) remains, however. Differences between monomer states in solutions of divalent and monovalent metals have apparently been ignored.

3.11. Interpretation of transport properties

The conductance data are comparable to that of electrolytes and one assumes ionization of the metal atom. The constancy of the Walden product (conductivity × viscosity) requires usual drift processes for the ion (Moynihan, Balitactac, Boone, and Litovitz 1971; Moynihan, Bressel, and Angell 1971). The temperature dependence almost matches that of the viscosity, but there is an unexplained excess of the Walden product (Dewald and Roberts 1968). One may separate out the individual equivalent conductances as shown in Table 3.3 (Berns 1965). An order-of-magnitude mobility for the solvated electron can be obtained from simple gas-kinetic considerations (Robinson and Stokes 1959; Lekner 1967; Cohen and Lekner 1967; Minday, Schmidt, and Davis 1971). Such considerations, of course, fail to account for the trends of cationic mobility within the alkali metal ions and do not contribute to the understanding of the solvated electron directly. However, the fact that rough agreement does exist between model and experiment indicates that no quantum mechanical (Grotthus) mechanisms are required, at least at low concentrations. Presumably the cavity deforms as an 'ameoba' (Kestner 1973). The thermoelectric power S and electrical conductivity σ are closely related (Cutler and Mott 1969; Cutler 1972; and Fritzsche 1971) whenever the electron carries all of its energy along with it–as for example an electron moving in the conduction band of a semiconductor carries its activation energy. Under this circumstance

$$d \ln \sigma / dS = -k_B / e, \qquad (3.28)$$

where k_B is Boltzmann's constant. Such a relation occurs for concentrated M–NH₃ solutions, but the slope is much less in dilute solutions (Thompson 1964). Much of the energy required to place the electron-in-a-cavity must therefore be left behind in the transport process. Lepoutre and Demortier (1971) have further pointed out that the large, negative heat-of-transfer of the electron can be well accounted for by the breakage of hydrogen bonds in the formation of the solvated electron (Kestner 1973). When the electron moves on, hydrogen bonds will reappear at the old site and be broken anew at the new site. Heat is transferred, then, in the direction opposite to the displacement of the electron. More on the mobility can be derived from magnetic resonance.

3.12. Interpretation of magnetic experiments

Indeed a great deal of information can be extracted from the resonance data (Catterall 1970). The data deserve a systematic treatment and one will be given. The exposition will proceed from the static nature of the various species to their dynamics.

The metal ion Knight shifts and relaxation times generally confirm the model of association embodied in eqns (3.19) and (3.20). The shift data are restricted (see Fig. 3.14) to the concentration range of appreciable ion pairing so that only the associated species are observed. The most dilute data indicate a rapid decrease in the shift as the concentration is decreased (O'Reilly 1964a), presumably because ion pairing is decreased (Demortier and Lepoutre 1969). O'Reilly has fitted his Knight shifts in the ion-pairing regime using the association process of eqn (3.19) together with the M^- species. In view of the discrepancies in equilibrium constants discussed previously, it does not seem worthwhile to put too much faith on the spin densities, etc. extracted by O'Reilly using eqn (3.5). They are nevertheless quoted here:

$$|\psi(\text{Li})|^2 = 6\cdot7 \times 10^{21} \text{ cm}^{-3}$$
$$|\psi(\text{Na})|^2 = 5\cdot4 \times 10^{22} \text{ cm}^{-3}$$
$$|\psi(\text{Rb})|^2 = 3\cdot4 \times 10^{23} \text{ cm}^{-3}$$
$$|\psi(\text{Cs})|^2 = 6\cdot8 \times 10^{23} \text{ cm}^{-3}$$

for $1\cdot0$ m.p.m. solutions (i.e. at the upper limit of the range of concentration considered in this chapter).

Demortier and Lepoutre (1969; Demortier 1972) conclude that their equilibrium constants based on eqn (3.19) and (3.20) are consistent with O'Reilly's data for Na–NH₃ solutions. They were, however, unable to obtain an exact fit, so Dye's (1970) dilemma persists (Catterall 1970). Despite the limitations of O'Reilly's values of the electronic spin density at the nucleus, one may use his results to estimate the lifetime of the species M. This computation (O'Reilly 1964a) leads to a lifetime near 10^{-12} seconds; that is, a time comparable to the reorientation time of the NH_3 molecule (Breitschwerdt and Radscheit 1969; O'Reilly 1969).

The discussion turns now to n.m.r. of the NH_3 molecule. The nitrogen magnetic resonance has been studied by O'Reilly (1964) and by Pinkowitz and Swift (1971). The total unpaired spin density at the ^{14}N nucleus is independent of the ion pairing process (Catterall 1970) and has a value $6.4 \times 10^{24} \, \text{cm}^{-3}$. The proton resonance is shifted to higher fields, an effect opposite to that seen at the ^{14}N nuclei and indicative of a lowered spin density at the proton. Catterall (1970) suggests the observed shift to be the superposition of a chemical shift due to hydrogen bond breaking in the formation of the electron cavity and a contact shift within the cavity. The negative spin density at the proton $(= -2.4 \times 10^{22} \, \text{cm}^{-3})$ has been attributed to a node of the solvated electron wavefunction at that point (O'Reilly 1964b) produced by orthogonalization with the molecular orbital wavefunctions. This explanation is, however, unconvincing in view of the solvated electron treatments by Copeland, Kestner, and Jortner (1970) and in view of similar shifts in other solvents (Catterall 1970) or even with transition metal nuclei (Wayland and Rice 1966).

Relaxation studies permit the calculation of the number of solvent molecules interacting with a solvated electron. The proton magnetic resonance (p.m.r.) spectra shows three peaks (Pinkowitz and Swift 1971); they arise from N–H coupling. The magnetic dipolar spin–lattice relaxation of the ^{14}N determines the shape of the p.m.r. spectra—in particular the relative heights and widths of the two outside peaks. Fitting parameters for the peaks include the associated relaxation time. In view of the rapid replacement of electron cosphere NH_3 molecules by bulk solvent (O'Reilly 1969), the ^{14}N spin–lattice relaxation time may be assigned to the solvated electron and the solvation number of the electron may be determined along with an improved estimate of the lifetime $\tau(e^-)$ of a given NH_3 in the neighbourhood of a solvated electron. (Note that Meiboom (1967) has criticized this technique as applied to aqueous solutions (Swift and Sayre 1966, 1967) on just the grounds that the exchange rate is too rapid to permit differentiation. There nevertheless seems to be agreement with hydration numbers obtained independently.) In the present case the electron resonance frequency could be varied with respect to $\tau(e^-)$ sufficiently to see an effect on the relaxation time. A solvation number typically between 20 and 40 was found, but no systematic variation with T or x could be detected within the uncertainties. Dwell times for the NH_3 were in the range $1-2 \times 10^{-12}$ s, again without discernable pattern. This time is near both orientational relaxation times (Breitschwerdt and Radscheit 1969) and diffusion times (O'Reilly 1969) so an assignment to one or the other process cannot be made.

The co-ordination number 20–40 greatly exceeds the values used by Copeland *et al.* (1970) or Feng *et al.* (1973), though it is close to O'Reilly's (1969) value. The rapid exchange must mean that solvent molecules well beyond the first co-ordination sphere contribute.

Catterall (1970) has also arrived at estimates of the co-ordination number based on proton relaxation. He suggests that both contact and dipolar relaxation processes are important. Only ratios involving the co-ordination number z, the proton–electron separation $R + a$ (see Fig. 3.37) and the dwell time for a solvent molecule $\tau(e^-)$ were obtained. They are

$$\tau(e^-)/z = 0{\cdot}6 \times 10^{-13} \text{ s},$$

$$z^2/(R + a)^6 = 3{\cdot}6 \times 10^{46} \text{ cm}^{-6}$$

The second figure may be compared with the result of Copeland *et al.* who find $z = 4$, $R = 1{\cdot}7 \times 10^{-8}$ cm (hard core radius of NH_3), so that $z^2/(R + a)^6 = 4 \times 10^{46}$ cm^{-6}. There is no real disagreement here, in contrast to the Pinkowitz and Swift result. Spin pairing does not change much (Feng, Fueki, and Kevan 1973).

Catterall's computation of $\tau(e^-)/z$ is based on electron relaxation times, as shown in Fig. 3.9. As cation effects have been ruled out, only electron–proton and electron–nitrogen interactions need be considered. As the proton Knight shift is so small only nitrogen effects need be considered (as already suggested by the effect obtained on replacing ^{14}N with ^{15}N (Pollack 1961), or protons by deuterons (O'Reilly 1961)). The contact interaction dominates in dilute solutions and

$$1/T_{le} \propto [z\,|\psi(N)|^2]\tau(e^-)/z \qquad (3.29)$$

where the factor in square brackets is evaluated from Knight shift data to be $6\cdot4\times10^{24}\,cm^{-3}$ and T_{le} is the electron spin–lattice relaxation time. The dwell time $\tau(e^-)$ to co-ordination number ratio is thus $0\cdot6\times10^{-13}\,s$, or if $z=4$ is used $\tau(e^-)=2\cdot4\times10^{-13}\,s$. As Catterall (1970) has pointed out this is in remarkable agreement with the correlation times for other processes in NH_3 (e.g. dielectric relaxation).

The assumption of T_{le} dominated by contact interaction with ^{14}N can only be justified in dilute solutions; at higher concentrations spin–orbit interactions enter. The differences between the various alkali metals (O'Reilly 1969) is attributed to increased spin–orbit interactions in the heavier solutes and, of course, requires the existence of ion-pairs. Chan, Austin, and Paez (1970) suggest, however, that the enhanced spin–orbit effects expected in Cs are not realized because of screening by the NH_3 molecules. Spin-pairing effects are small and the lifetime of the spin-paired species must exceed $10^{-6}\,s$.

In sum, the magnetic data support the model of the solvated electron proposed by Copeland $et\ al.$ (1970) (see also Catterall and Mott 1969) together with the association picture which considers loose aggregates of ions to dominate (Gold, Jolly, and Pitzer 1962; Cohen and Thompson 1968; Dye 1970). Some numerical estimates of spin densities are available to keep the theorists honest. A regrettably small amount of detail has emerged concerning the spin-paired species as it does not contribute to the resonance process.

3.13. Thermodynamic considerations

Thermodynamic arguments have also been employed in attempting to understand the many measurements. Consideration of heats of ammoniation for the solvated electron provides yet another check of the energetics of the solvation process as described by Jortner and others. Activities and entropies also may be obtained and hold potential clues to the nature of the species present in these unique materials. The subject has been reviewed by Coulter (1953) as well as in the general reviews (e.g. Das 1962).

The heat of ammoniation of the electron is extracted from the measured heat of solution (Gunn 1964) by assuming a cycle of the form: sublimation of solid metal, ionization of metal vapour, ammoniation of metal ion, and finally ammoniation of electron (Jortner 1959; Lepoutre and Demortier 1971; Lepoutre and Jortner 1972). The first three terms can be estimated in accord with standard chemical practice (Das 1962) and the known heat of solution permits calculation of the final term. There is no dependence on solute, including divalent metals, and the result is $1·1 \pm 0·1$ eV. This compares well with the $1·0$ eV obtained by Copeland, Kestner, and Jortner (1970) using $V_0 = +0·5$ eV. Land and O'Reilly (1967) originally obtained $3·4$ eV but O'Reilly (1971) now reports a satisfactory value of $1·5$ eV. Feng et al. (1973) find $1·9$ eV. Comparisons require sufficient assumptions as to weaken any notion of success.

The change in the heat of solution at higher concentration is $50 \, \text{cal} \, \text{mol}^{-1} \, \text{deg}^{-1}$ (Gunn 1963) which is equivalent to $-0·2 \pm 0·05$ eV as the energy change in the spin pairing process, that is equivalent to an energy gain near to $0·1$ eV per electron. This is also consistent with the activation energy for spin unpairing observed by Hutchison and Pastor (1953), and with the ammoniation energies calculated by Russell and Sienko (1957) from e.m.f.s in galvanic cells including Na–NH₃ solutions. These numbers are also quite close to the change in the energy of the optical absorption maximum, see Fig. 3.19. O'Reilly (1971) has recently obtained $0·19$ eV for the gain in spin pairing; Kestner and Copeland (1970) obtain $0·9$ eV. Feng et al. (1973) get $0·65$ eV. Correlation energy estimates lead to considerable uncertainties.

Activities (eqn 3.15) may be computed from vapour pressure data (Marshall 1962, 1964; Dye, Lepoutre, Marshall, and Pajot 1964). The calculation requires the definition of a reference state which is typically chosen so that $\gamma_{Na}k = 1·0$ at a concentration of 1 molal [moles solute/kg solvent]. The value of the adjustable constants k is then fixed by using some alternate method to fix γ_{Na} (Dye, Lepoutre, Marshall, and Pajot 1964). Ichikawa and Thompson's (1973) recent results using cell e.m.f.s in Na–NH₃

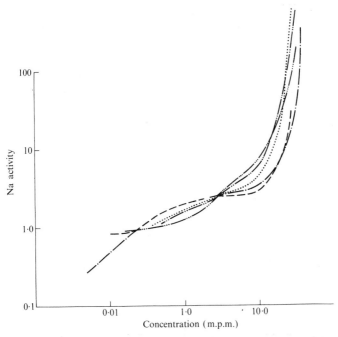

FIG. 3.39. Activities for several metal cations in M–NH₃ solutions (Marshall 1962).

solutions are consistent with the vapour pressure data over most of the concentration range. For consistency the vapour pressure data were used to compute the activities for several M–NH₃ solutions as shown in Fig. 3.39; the tendency toward a plateau at intermediate concentrations is indicative of the phase transition seen there (Chapter 5). A more detailed view of the available Na–NH₃ data is shown in Fig. 2.20. That figure is based not only on the vapour pressure data but contains, as well, activities calculated from e.m.f. and transference number data. Use of the extended Debye–Hückel theory (Harned and Owen 1958) together with an assumed 'distance of closest approach' for the ion pairs permits one to once again check the equilibrium constants of eqn (3.19) and (3.20). Values close to those of Demortier and Lepoutre (1969, 1972) are obtained.

Further thermodynamic analysis can be based on the decomposition reaction [eqn 3.2] which may be reversed upon the application of pressure (Kirschke and Jolly 1967). That equilibrium together with the dissolution reaction and the temperature dependencies of the various equilibria leads to enthalpies and free energies of formation of the solvated electron and, finally, to the entropy of the solvated electron (Lepoutre and Demortier 1971). The final result is $37 \pm 5 \cdot 4$ cal mol^{-1} deg^{-1} at 238 K, when the zero of entropy is absolute. This value compares well, both as to sign and magnitude, with that based on the thermoelectric entropy of transport (Dewald and Lepoutre 1956; Lepoutre and Demortier 1971). Entropies of the other constituents of (somewhat decomposed) M–NH$_3$ solutions are as follows:

Ion	Entropy at 238 K
H$^+$	-28 cal mol^{-1} deg^{-1}
NH$_2^-$	$8 \cdot 1$
Cl$^-$	$-3 \cdot 9$
K$^+$	$-2 \cdot 6$
Ca^{2+}	-185
e$^-$	37
NH$_3$ (liq.)	$20 \cdot 8$
NH$_3$ (gas)	44

Clearly those ions that tightly bind solvent molecules (H$^+$ or Ca^{2+}) reduce the entropy of the system. The effect of the solvated electron appears to be to disorganize the solvent in the sense of breaking hydrogen bonds. At higher temperatures the e$^-$ entropy is reduced to 18 cal mol^{-1} deg^{-1} (Kirschke and Jolly 1967), the excess volume is greatly reduced (Naiditch, Paez, and Thompson 1967) and the heat of solution is also smaller (Gunn 1964). Thermal agitation has already disordered the fluid and the electron has less to do. The Catterall–Mott (1969) view of solvation, seen in Fig. 3.33, suggests that the molecules in the first co-ordination layer are pushed apart, thereby breaking hydrogen bonds (Kestner 1973). At higher temperatures the bonds are already distorted, and the average separation of the NH$_3$ molecules is greater (the free volume is greater). The electron therefore perturbs the solvent less.

3.14. Summary

We are finally in a position to summarize the current description of solvated electrons and of dilute metal–ammonia solutions.

The picture is essentially that of Fig. 3.33 where the electron cavity is shown to consist of a solvent vacancy surrounded by a co-ordination shell containing 4 to 8 solvent molecules at greater-than-usual separation. The exchange of cosphere and bulk solvent is exceedingly rapid, so that the identity of the cavity 'wall' is lost while the 'wall' retains its integrity. The solute cation is also solvated and is located somewhere else. The potential well formed by the cavity together with bulk solvent polarization is sufficient to bind the electron and to provide one or more bound excited states. At the lowest concentrations the solvated electron moves in response to an applied field as a more or less rigid entity though there is evidence that reorientation of the cosphere molecules permits a somewhat greater mobility than Stoke's law would indicate.

As the concentration is increased there is association of cations and electrons first into pairs then spin-paired species. The most economical constructs are electrostatically bound clusters of 2, 4, or more charges. Spin-pairing may put two electrons into a single cavity. Though little has been made of it as yet, larger and larger clusters must form as the concentration increases (Cohen and Thompson 1968; Chieux and Sienko 1970; Jortner and Cohen 1973; and Lelieur 1973). Nevertheless, the structure of the cavity and the energy of the electron is largely unaffected by the association. Impressions that the spin-paired species deserves more study are growing (Kestner 1973; Feng *et al.* 1973; Rubinstein, Tuttle, and Golden 1973) and the old experiments will not suffice. One needs to determine not only the optical band and heat of solution change when spin-pairing occurs but the dissociation energy of the species containing the two electrons and the separation of singlet and triplet states. Since spin-pairing occurs at lower concentrations in $Eu–NH_3$ solutions (Catterall and Symons 1965) detailed studies must be made of cation effects using both alkali and alkaline earth solutes. It seems likely that cation effects will appear in *careful* experiments involving almost any property sensitive to association processes despite the general assertion that cations are *relatively* unimportant.

There are frustrating quantitative inconsistencies amongst the association parameters. It seems preferable at present to ascribe these to experimental problems or conceptual difficulties within the theory of associated electrolytes (Allnatt and Cohen 1964; Frank 1966) rather than to modify the picture presented above.

The association processes produce notable but not dramatic changes in the properties of the solutions. We turn in the next chapters to dramatic effects resulting from the metal–nonmetal transition.

REFERENCES

ACRIVOS, J. V. and PITZER, K. S. (1962). *J. phys. Chem.* **66,** 1693.

ALGER, R. S. (1968). *Electron paramagnetic resonance.* p. 290 *et seq.* Interscience, New York.

ALLNATT, A. R. and COHEN, M. H. (1964*a*). *J. chem. Phys.* **40,** 1860.

—— —— (1964*b*). *J. chem. Phys.* **40,** 1871.

ANDREAE, J. H., EDMONDS, P. D., and McKELLAR, J. F. (1965). *Acustica* **15,** 74.

ANGELL, C. A. and SARE, E. J. (1968). *J. chem. Phys.* **49,** 4713.

APFEL, H., BREITSCHWERDT, K., and SCHMIDT, W. (1968). *Ber.* (*Dtsch*) *Bunsenges. phys. Chem.* **72,** 1059.

ARNOLD, E. and PATTERSON, A. (1964*a*). *J. chem. Phys.* **41,** 3089.

—— —— (1964*b*). *J. chem. Phys.* **41,** 3098.

AULICH, H., BARON, B., DELAHAY, P., and LUGO, R. (1973). *J. chem. Phys.* **58,** 4439.

BAILEY, K. E. and BOWEN, D. E. (1972). *J. chem. Phys.* **56,** 4809.

BARON, B., CHARTIER, P., DELAHAY, P., and LUGO, R. (1969). *J. chem. Phys.* **51,** 2562.

—— DELAHAY, P., and LUGO, R. (1970). *J. chem. Phys.* **53,** 1399.

——, ——, —— (1971). *J. chem. Phys.* **55,** 4180.

BECKER, E., LINDQUIST, R. H., and ALDER, B. J. (1956). *J. chem. Phys.* **25,** 971.

BECKMAN, T. A. and PITZER, K. S. (1961). *J. phys. Chem.* **65,** 1527.

BELLISSENT, M. C., GÉRARD, P., LONGVIALLE, C., METON, M., PICK, M., and MORAND, G. (1971). *J. Chim. phys.* **68,** 355.

BELLONI, J. and FRADIN DE LA RENAUDIERE, J. (1971). *Nature Lond.* **232,** 173.

BERNS, D. S. (1964). In *Metal–ammonia solutions* (eds G. Lepoutre and M. J. Sienko), p. 146. Benjamin, New York.

—— (1965). *Adv. Chem. Ser.* **50,** 82.

—— LEPOUTRE, G., BOCKELMAN, E. A., and PATTERSON, A. (1961). *J. chem. Phys.* **35,** 1820 (1961).

BIRCHALL, T. and JOLLY, W. L. (1965). *J. Am. chem. Soc.* **87,** 3007.

BJERRUM, N. (1926). *K. danske Vidensk. Selsk. Skr.* **7,**

BLANDER, M. (1964). In *Molten salt chemistry* (ed. M. Blander). Interscience, New York.

BÖDDEKER, K. W., LANG, G., and SCHINDEWOLF, U. (1970). In *Metal–ammonia solutions* (eds J. J. Lagowski and M. J. Sienko), p. 219. Butterworths, London.

—— and VOGELGESANG, R. (1971). *Ber.* (*Dtsch*) *Bunsenges. phys. Chem.* **75,** 638.

BOWEN, D. E. (1969). *J. chem. Phys.* **51,** 1115.

—— (1970). In *Metal-ammonia solutions* (eds J. J. Lagowski and M. J. Sienko), p. 355. Butterworths, London.

—— PARKER, M. H. and FRANCEWARE, L. B. (1973). *J. chem. Phys.* **58,** 2984.

—— THOMPSON, J. C., and MILLETT, W. E. (1968). *Phys. Rev.* **168,** 114.

BRADY, G. W. and VARIMBI, J. (1964). *J. chem. Phys.* **40,** 2615.

BREITSCHWERDT, K. G. and RADSCHEIT, H. (1969). *Phys. Letts* **29A**, 381.
—— —— (1971). *Ber. (Dtsch) Bunsenges. phys. Chem.* **75**, 644.
—— —— (1973). In *Electrons in fluids* (eds J. Jortner and N. R. Kestner), p. 315. Springer-Verlag, Heidelberg.
—— —— and WOLZ, H. (1974). In *Amorphous and liquid semiconductors* (eds J. Stuke and W. Brenig), p. 1337. Taylor and Francis, London.
—— and SCHMIDT, W. (1970). *Z. Naturf.* **25A**, 1467.
BRENDLEY, W. H. and EVERS, E. C. (1965). *Adv. Chem. Ser.* **50**, 111.
BRIDGES, R., INGLE, A. J., and BOWEN, D. E. (1970). *J. chem. Phys.* **52**, 5106.
BRONSKILL, M. J., WOLFF, R. K., and HUNT, J. W. (1969). *J. phys. Chem.* **73**, 1175.
BUROW, D. F. and LAGOWSKI, J. J. (1965). *Adv. Chem. Ser.* **50**, 125.
CARVER, T. R. and SLICHTER, C. P. (1956). *Phys. Rev.* **102**, 975.
CATTERALL, R. (1970). *Phil. Mag.* **22**, 779.
—— and MOTT, N. F. (1969). *Adv. Phys.* **18**, 665.
—— and SYMONS, M. C. R. (1964). *J. chem. Soc.* **1964**, 4342.
—— —— (1965). *J. chem. Phys.* **42**, 1466.
CHAN, S. I., AUSTIN, J. A., and PAEZ, O. A. (1970). In *Metal–ammonia solutions* (eds J. J. Lagowski and M. J. Sienko), p. 425. Butterworths, London.
CHIEUX, P. and SIENKO, M. J. (1970). *J. chem. Phys.* **53**, 566.
CLARK, H. C., HORSFIELD, A., and SYMONS, M. C. R. (1959). *J. chem. Soc.* **1959**, 2478.
COHAN, N. V., FINKELSTEIN, G., and WEISSMAN, M. (1974). *Chem. Phys. Letts* **26**, 93.
COHEN, M. H. and LEKNER, J. (1967). *Phys. Rev.* **158**, 305.
—— and THOMPSON, J. C. (1968). *Adv. Phys.* **17**, 857.
CONWAY, B. E. and BARRADAS, R. G. (1966). *Chemical physics of ionic solutions.* Wiley, New York.
COPELAND, D. A., KESTNER, N. R., and JORTNER, J. (1970). *J. chem. Phys.* **53**, 1189.
CORSET, J., HOUNG, P. F., and LASCOMBE, J. (1968). *Spectrochim. Acta* **24A**, 1385.
COULTER, L. V. (1953). *J. phys. Chem.* **57**, 553.
CUTLER, M. (1972). *Phil. Mag.* **25**, 173.
—— and MOTT, N. F. (1969). *Phys. Rev.* **181**, 1336.
DANNER, J. C. and TUTTLE, T. R. (1963). *J. Am. chem. Soc.* **85**, 4052.
DAS, T. P. (1962). *Adv. chem. Phys.* **4**, 303.
DAVIDSON, N. (1962). *Statistical mechanics.* McGraw-Hill, New York.
DE GROOT, S. R. (1959). *Thermodynamics of irreversible processes.* North-Holland, Amsterdam.
DELAHAY, P. (1971). *J. chem. Phys.* **55**, 4188.
DELBECQ, C. J., PRINGSHEIM, P., and YUSTER, P. (1951). *J. chem. Phys.* **19**, 574.
——, ——, —— (1952). *J. chem. Phys.* **20**, 746.
DEMORTIER, A. (1970). thesis, Lille. (Unpublished).
—— DEBACKER, M., and LEPOUTRE, G. (1972). *J. Chim. phys.* **69**, 380.
—— and LEPOUTRE, G. (1969). *C. r. hebd. Sceanc. Acad. Sci. Paris* **268**, 453.
—— (1972). *J. Chim. phys.* **69**, 179.
—— LOBRY, P., and LEPOUTRE, G. (1971). *J. Chim. phys.* **68**, 498.
DEWALD, J. F. and LEPOUTRE, G. (1954). *J. Am. chem. Soc.* **76**, 3369.
—— —— (1956). *J. Am. chem. Soc.* **78**, 2956.
DEWALD, R. R. (1969). *J. phys. Chem.* **23**, 2615.
—— and ROBERTS, J. H. (1968). *J. phys. Chem.* **72**, 4224.
—— and TSINA, R. V. (1968). *J. phys. Chem.* **72**, 4520.
DEXTER, D. L. and FOWLER, W. B. (1969). *Phys. Rev.* **183**, 307.

DYE, J. L. (1970). In *Metal–ammonia solutions* (eds. J. J. Lagowski and M. J. Sienko), p. 1. Butterworths, London.

—— DEBACKER, M. G., and DORFMAN, L. (1970). *J. chem. Phys.* **52**, 6251.

DYE, J. L., LEPOUTRE, G., MARSHALL, P. R., and PAJOT, P. (1964). In *Metal–ammonia solutions* (eds G. Lepoutre and M. J. Sienko), p. 92. Benjamin, New York.

——, SANKUER, R. F., and SMITH, G. E. (1960). *J. Am. chem. Soc.* **82**, 4797.

——, SMITH, G. E., and SANKUER, R. F. (1960). *J. Am. chem. Soc.* **82**, 4803.

EGGARTER, T. P., and COHEN, M. H. (1970). *Phys. Letts* **25**, 807.

FENG, D.-F., FUEKI, K., and KEVAN, L. (1973). *J. chem. Phys.* **58**, 3281.

FRANCEWARE, L. B., PRIESAND, M. A., and BOWEN, D. E. (1972). *J. chem. Phys.* **57**, 4099.

FRANK, H. S. (1966). In *Chemical physics of ionic solutions* (eds B. E. Conway and R. G. Barradas), p. 53. Wiley, New York.

FRANKLIN, E. C. (1910). *Z. phys. Chem.* **69**, 272.

FRITZSCHE, H. (1971). *Solid St. Commun.* **9**, 1813.

FUEKI, K. (1969). *J. chem. Phys.* **50**, 5381.

—— and NODA, S. (1970). In *Metal–ammonia Solutions* (eds J. J. Lagowski and M. J. Sienko), p. 19. Butterworths, London.

GARNSEY, R., BOE, R. J., MAHONEY, R., and LITOVITZ, T. A. (1969). *J. chem. Phys.* **50**, 5222.

GOLD, M. and JOLLY, W. L. (1962). *Inorg. Chem.* **1**, 818.

——, ——, and PITZER, K. S. (1962). *J. Am. chem. Soc.* **84**, 2264.

GOLDEN, S., GUTTMAN, C., and TUTTLE, T. R. (1965). *J. Am. chem. Soc.* **87**, 135.

——, ——, —— (1966). *J. chem. Phys.* **44**, 3791.

GORDON, R. and SUNDHEIM, B. (1964). *J. phys. Chem.* **68**, 3347.

GRANTHAM, L. F. and YOSIM, S. J. (1966). *J. chem. Phys.* **45**, 1192.

—— —— (1963). *J. chem. Phys.* **38**, 1671.

GREEN, J. H. and LEE, J. (1964). *Positronium Chemistry.* Academic Press, New York.

GUNN, S. R. (1964). In *Metal–ammonia solutions* (eds G. Lepoutre and M. J. Sienko), p. 76. Benjamin, New York.

—— (1967). *J. chem. Phys.* **47**, 1174.

—— and GREEN, L. G. (1962). *J. chem. Phys.* **36**, 368.

HARNED, H. S. and OWEN, B. B. (1958). *Physical chemistry of electrolytic solutions.* Reinhold, New York.

HART, E. J. (1970). In *Metal–ammonia solutions* (eds J. J. Lagowski and M. J. Sienko), p. 413. Butterworths, London.

HÄSING, J. (1940). *Annln Phys.* **37**, 509.

HASTED, J. B., RITSON, D. M., and COLLIE, C. H. (1948). *J. chem. Phys.* **16**, 1.

—— and TIRMAZI, S. H. (1969). *J. chem. Phys.* **50**, 4116.

HILL, T. L. (1948). *J. chem. Phys.* **16**, 394.

HNIZDA, V. F. and KRAUS, C. A. (1949). *J. Am. chem. Soc.* **71**, 1565.

HUGHES, T. R. (1963). *J. chem. Phys.* **38**, 202.

HUTCHISON, C. A. (1953). *J. phys. Chem.* **57**, 546.

—— and O'REILLY, D. E. (1961). *J. chem. Phys.* **34**, 163.

—— —— (1970). *J. chem. Phys.* **52**, 4400.

—— and PASTOR, R. (1953a). *Rev. mod. Phys.* **25**, 285.

—— —— (1953b). *J. chem. Phys.* **21**, 1959.

ICHIKAWA, K. and SHIMOJI, M. (1967). *Ber. (Dtsch) Bunsenges. phys. Chem.* **71**, 1149.

—— —— (1969). *Ber. (Dtsch) Bunsenges. Phys. Chem.* **73**, 302.

—— and THOMPSON, J. C. (1973). *J. chem. Phys.* **59**, 1680.

IGUCHI, K. (1968). *J. chem. Phys.* **48**, 1735.

ITOH, J. and TAKEDA, T. (1963). *J. phys. Soc. Japan* **18**, 1560.
JAHNKE, J. A., MEYER, L., and RICE, S. A. (1971). *Phys. Rev.* **A3**, 734.
JOHNSON, W. C. and MEYER, A. W. (1932). *J. Am. chem. Soc.* **54**, 3621.
JOLLY, W. L. (1959). *Prog. inorg. Chem.* **1**, 235.
JORTNER, J. (1959). *J. chem. Phys.* **30**, 839.
—— and COHEN, M. H. (1973). *J. chem. Phys.* **58**, 5170.
—— and KESTNER, N. R. (1970). In *Metal–ammonia solutions* (eds J. J. Lagowski and M. J. Sienko), p. 49. Butterworths, London.
——, RICE, S. A., and WILSON, E. G. (1964). *Metal–ammonia solutions* (eds G. Lepoutre and M. J. Sienko), p. 222. Benjamin, New York.
KAJIWARA, T., FUNABASHI, K., and NALEWAY, C. (1972). *Phys. Rev. A* **6**, 808.
KAPLAN, J. and KITTEL, C. (1953). *J. chem. Phys.* **21**, 1429.
KAY, R. L. and DYE, J. L. (1963). *Proc. Natn. Acad Sci. N.Y.* **49**, 5.
KENESHEA, F. J., JR. and CUBICCIOTTI, D. (1958). *J. phys. Chem.* **62**, 843.
KESTNER, N. R. (1973). In *Electrons in fluids* (eds J. Jortner and N. R. Kestner), p. 1. Springer-Verlag, Heidelberg.
—— and JORTNER, J. (1973). *J. phys. Chem.* **77**, 1040.
——, ——, and GAATHON, A. (1973). *Chem. Phys. Lett.* **19**, 328.
KIKUCHI, S. (1939). *J. Soc. chem. Ind. Japan* **42** (Suppl. Binding) 15.
—— (1944). *J. Soc. chem. Ind. Japan* **47**, 488.
KIRSCHKE, E. J. and JOLLY, W. L. (1967). *Inorg. Chem.* **6**, 885.
KOEHLER, W. H. and LAGOWSKI, J. J. (1969). *J. phys. Chem.* **73**, 2329.
KRAUS, C. A. (1908). *J. Am. chem. Soc.* **30**, 1197.
—— (1914). *J. Am. chem. Soc.* **36**, 864.
—— (1921*a*). *J. Am. chem. Soc.* **43**, 741.
—— (1921*b*). *J. Am. chem. Soc.* **43**, 749.
——, CARNEY, E. S., and JOHNSON, W. C. (1927). *J. Am. chem. Soc.* **49**, 2206.
—— and LUCASSE, W. W. (1921). *J. Am. chem. Soc.* **43**, 2529.
LAMBERT, C. (1968). *J. chem. Phys.* **48**, 2389.
LAND, R. H. and O'REILLY (1967). *J. chem. Phys.* **46**, 4496.
LEKNER, J. (1967). *Phys. Rev.* **158**, 130.
LELIEUR, J.-P. (1973). *J. chem. Phys.* **59**, 3510.
LEPOUTRE, G. and DEMORTIER, A. (1971). *Ber. (Dtsch) Bunsenges. phys. Chem.* **75**, 647.
—— and DEWALD, J. F. (1956). *J. Am. chem. Soc.* **78**, 2956.
—— and JORTNER, J. (1972). *J. phys. Chem.* **76**, 683.
—— and LELIEUR, J.-P. (1970). In *Metal–ammonia solutions* (eds J. J. Lagowski and M. J. Sienko), p. 247. Butterworths, London.
LIPSCOMB, W. N. (1953). *J. chem. Phys.* **21**, 52.
LOBRY, P. (1969). Thesis, Lille. (Unpublished).
MAHAFFEY, D. W. and JERDE, D. A. (1968). *Rev. mod. Phys.* **40**, 710.
MARSHALL, P. R. (1962). *J. chem. Engng Data* **7**, 399.
—— (1964). In *Metal–ammonia Solutions* (eds G. Lepoutre and M. J. Sienko), p. 97. Benjamin, New York.
MAYBURY, R. H. and COULTER, L. V. (1951). *J. chem. Phys.* **19**, 1326.
MCCONNEL, H. M. and HOLM, C. H. (1957). *J. chem. Phys.* **26**, 1517.
MECHLIN, G. (1952). Thesis, University of Pittsburgh. (Unpublished).
MEIBOOM, S. (1967). *J. chem. Phys.* **46**, 410.
MEYER, L., DAVIS, H. T., RICE, S. A., and DONNELLY, R. J. (1962). *Phys. Rev.* **126**, 1927.
MICHAEL, B. D., HART, E. J., and SCHMIDT, K. H. (1970). *J. phys. Chem.* **76**, 2798.
MINDAY, R. M., SCHMIDT, L. D., and DAVIS, H. T. (1971). *J. chem. Phys.* **54**, 3112.

MORALES, R. and MILLETT, W. E. (1967). *Bull. Am. phys. Soc.* **12**, 193.
MOYNIHAN, C. T., BALITACTAC, N., BOONE, L., and LITOVITZ, T. A. (1971). *J. chem. Phys.* **55**, 3013.
—— BRESSEL, R. D., and ANGELL, C. A. (1971). *J. chem. Phys.* **55**, 4414.
MUELLER, W. E. and THOMPSON, J. C. (1970). In *Metal–ammonia solutions* (eds J. J. Lagowski and M. J. Sienko), p. 293. Butterworths, London.
NAIDITCH, S., PAEZ, O. A., and THOMPSON, J. C. (1967). *J. chem. Engng Data* **12**, 164.
NAKAMURA, Y., HORIE, Y., and SHIMOJI, M. (1974). *Discuss. Faraday Soc. I* **70**, 1376.
NASBY, R. D. and THOMPSON, J. C. (1970). *J. chem. Phys.* **53**, 109.
NEWMARK, R. A., STEPHENSON, J. C., and WAUGH, J. S. (1967). *J. chem. Phys.* **46**, 3514.
NEWTON, M. (1973). *J. chem. Phys.* **58**, 5833.
NOZAKI, T. and SHIMOJI, M. (1969). *Trans. Faraday Soc.* **65**, 1489.
OGG, R. A. (1946a). *Phys. Rev.* **69**, 243.
—— (1946b). *Phys. Rev.* **70**, 93.
—— (1946c). *J. Am. chem. Soc.* **68**, 155.
—— (1946d). *J. chem. Phys.* **14**, 295.
ONSAGER, L. (1968). *Rev. mod. Phys.* **40**, 709.
O'REILLY, D. E. (1961). *J. chem. Phys.* **34**, 1279.
—— (1963). *Phys. Rev. Letts* **11**, 545.
—— (1964a). *J. chem. Phys.* **41**, 3729.
—— (1964b). *J. chem. Phys.* **41**, 3736.
—— (1969a). *J. chem. Phys.* **50**, 4320.
—— (1969b). *J. chem. Phys.* **50**, 4743.
—— (1969c). *J. chem. Phys.* **50**, 5378.
—— (1971). *J. chem. Phys.* **55**, 474.
OVERHAUSER, A. W. (1953). *Phys. Rev.* **92**, 411.
PARKER, M. H. and BOWEN, D. E. (1971). *Ber. (Dtsch) Bunsenges. phys. Chem.* **75**, 639.
PINKOWITZ, R. A. and SWIFT, T. J. (1971). *J. chem. Phys.* **54**, 2858.
PITZER, K. (1964). In *Metal–ammonia solutions* (eds G. Lepoutre and M. J. Sienko), p. 193. Benjamin, New York.
POLLAK, V. L. (1961). *J. chem. Phys.* **34**, 864.
POTTER, R. L., SHORES, R. G., and DYE, J. L. (1961). *J. chem. Phys.* **35**, 1907.
QUINN, R. K. and LAGOWSKI, J. J. (1968). *J. phys. Chem.* **72**, 1374.
RALEIGH, D. O. (1963). *J. chem. Phys.* **38**, 1677.
RAMANATHAN, P. S. and FRIEDMAN, H. L. (1971). *J. chem. Phys.* **54**, 1086.
RENTZEPIS, P. M., JONES, R. P., and JORTNER, J. (1973). *J. chem. Phys.* **59**, 766.
ROBINSON, R. A. and STOKES, R. H. (1959). *Theory of electrolyte solutions* (2nd edn). Butterworths, London.
RUBINSTEIN, G., TUTTLE, T. R., and GOLDEN, S. (1973). *J. phys. Chem.* **77**, 2872.
RUSCH, P. F., KOEHLER, W. H., and LAGOWSKI, J. J. (1970). In *Metal–ammonia solutions* (eds J. J. Lagowski and M. J. Sienko), p. 41. Butterworths, London.
—— and LAGOWSKI, J. J. (1973). *J. phys. Chem.* **77**, 210.
RUSSELL, J. B. and SIENKO, M. J. (1957). *J. Am. chem. Soc.* **79**, 4051.
SAMOILOV, O. YA. (1957). *Discuss. Faraday Soc.* **24**, 141.
SCHETTLER, P. D. and PATTERSON, A. (1970). In *Metal–ammonia solutions* (eds J. J. Lagowski and M. J. Sienko), p. 395. Butterworths, London.
——, VAN ANTWERP, C. L., HAMILTON, J. A., THILLY, J. E., and SPEAR, J. D. (1973). In *Electrons in fluids* (eds J. Jortner and N. R. Kestner), p. 239. Springer-Verlag, Heidelberg.

SCHMEIG, G. M. (1939). Dissertation, Chicago. (Unpublished).

SCHMIDT, P. W. (1957). *J. chem. Phys.* **27**, 23.

SIENKO, M. J. (1964). In *Metal–ammonia solutions* (eds G. Lepoutre and M. J. Sienko), p. 23. Benjamin, New York.

SOLOMON, I. and BLOEMBERGEN, N. (1956). *J. chem. Phys.* **25**, 261.

SPRINGETT, B. E., JORTNER, J., and COHEN, M. H. (1968). *J. chem. Phys.* **48**, 2720.

STAIRS, R. A. and SIENKO, M. J. (1956). *J. Am. chem. Soc.* **78**, 920.

STILLINGER, F. H. and LOVETT, R. (1968*a*). *J. chem. Phys.* **48**, 3858.

—— —— (1968*b*). *J. chem. Phys.* **48**, 3869.

SUNDHEIM, B. R. (1964) (ed.). *Fused salts*, p. 165. McGraw-Hill, New York.

SWIFT, T. J. and LO, H. H. (1967). *J. Am. chem. Soc.* **89**, 3988.

—— MARKS, S. B., and SAYRE, W. G. (1966). *J. chem. Phys.* **44**, 2797.

—— and SAYRE, W. G. (1966). *J. chem. Phys.* **44**, 3567.

—— —— (1967). *J. chem. Phys.* **46**, 410.

TEAL, G. (1948). *Phys. Rev.* **71**, 138.

THOMPSON, J. C. (1964). In *Metal–ammonia solutions* (eds. G. Lepoutre and M. J. Sienko), p. 307. Benjamin, New York.

—— (1967). In *Chemistry of non-aqueous solvents* (ed. J. J. Lagowski), p. 265. Academic Press, New York.

—— and ORÉ-ORÉ, C. R. (1971). *J. chem. Phys.* **54**, 2279.

TOPOL, L. E. and RANSOM, L. D. (1963). *J. chem. Phys.* **38**, 1663.

TUTTLE, T. R., GUTTMAN, C., and GOLDEN, S. (1966). *J. chem. Phys.* **45**, 2206.

VOGRIN, F. J., KNAPP, P. S., FLINT, W. L., ANTON, A., HIGHBERGER, G., and MALINKOWSKI, E. R. (1971). *J. chem. Phys.* **54**, 178.

WALL, T. T. and HORNIG, D. F. (1967). *J. chem. Phys.* **47**, 784.

WAYLAND, B. B. and RICE, W. L. (1966). *J. chem. Phys.* **45**, 3150.

WEST, R. N. (1973). *Adv. Phys.* **22**, 263.

4

THE METAL–NONMETAL
TRANSITION IN METAL–
AMMONIA SOLUTIONS

4.1. Introduction

IN previous chapters dilute solutions of alkali metals in liquid ammonia have been shown to behave in a manner analogous to that of dilute solutions of salts in liquid ammonia, whereas concentrated solutions of alkali metals in liquid ammonia behave quite similarly to liquid metals. Between the dilute and concentrated ranges there should be a transitional range in which the behaviour of the solutions changes from that comparable to an electrolyte to a behaviour comparable to a liquid metal. Indeed this transition will be shown to occur over a fairly narrow concentration range and may be compared to what has been called by N. F. Mott a metal–nonmetal transition. As the ideas and nomenclature of the metal–nonmetal transition are newer and have not become as familiar as have the ideas involved in the free-electron theory of simple metals or in the Debye–Hückel theory of electrolytes, this chapter begins with a brief discussion of one of the mechanisms possibly responsible for a metal-nonmetal (M–NM) transition so that the data presentation may appear more reasonable.

The metallic state is characterized by electron wavefunctions which are delocalized and which, in the solid state at least, may be described as Bloch waves or in the simplest metals as plane waves. In contrast, in the nonmetallic state the electron wavefunction is localized, described perhaps by a Wannier function having an appreciable amplitude only over a volume of space small compared to the volume of the system. One of the earliest examples of a M–NM transition, showing this transition from plane waves to localized wavefunctions, was that of the electron crystal, as discussed by Wigner in the thirties. Wigner (1938) considered a model which contained a dispersion of electrons within a homogeneous, continuous, positively-charged background. When the electron density is high, screening effects reduce the Coulomb interaction of the electrons so that the kinetic energy may dominate and the electrons may be accurately described in terms of plane waves. As the electron density is reduced, screening becomes less effective and the Coulomb repulsion between the electrons forces them into a structure wherein the electron wavefunctions are localized on the points of a lattice (not the lattice of ions). When the electrons are localized on lattice points, they no longer may move under the influence of an applied electric field and the system may *not* be described as a conductor. Wigner thus described a metal–insulator transition. One may expect to see in such a system an abrupt change in the dependence of the conductivity upon the interelectronic distance when the screening is no longer effective; one may expect also to see rather abrupt changes in the free energy as one goes from a system where Coulomb forces are dominant to a system where the translational kinetic energy is dominant. The breadth of the M–NM transition might be expected to increase at temperatures so high that the Fermi system is no longer completely degenerate and would also increase as a result of thermal excitations.

Though the electron crystal is now known not to be a useful approximation to a low density assembly of one electron centres there are, nevertheless, many other models which lead to localization of the electron wavefunction (Mott and Zinamon 1970; Mott 1974c). The idea that electron–electron interactions, antiferromagnetism, disorder, or other forces may lead to the localization of the electron wavefunction will provide the framework within which the data on the metal–ammonia solutions will be discussed below. Even this brief discussion suggests that the identification of an M–NM transition be based primarily on the conductivity.

4.2. M–NH₃ data

Reviews have been given by Thompson (1968), Cohen and Thompson (1968), Lepoutre and Lelieur (1970), and Lepoutre (1973).

Fig. 4.1 illustrates the change of the conductivity with concentration in lithium–ammonia solutions (Nasby and Thompson

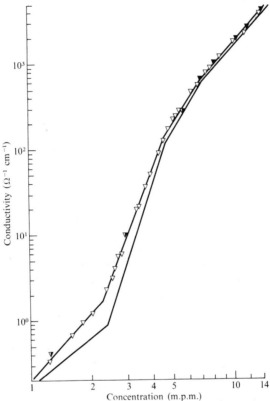

FIG. 4.1. Conductivity versus concentration. Data are shown at 234 K, and the
line is for 211 K (Nasby and Thompson 1970). See the caption to Fig. 2.1 (p. 15)
for the notation in this and subsequent figures.

1970). In that figure the conductivity is a fairly strong function of
the concentration up to approximately 2 m.p.m. whereupon it
rises even more sharply with concentration until the concentration reaches 6 m.p.m. above which concentration it shows a
somewhat more gentle rise. In the steepest portion of the curve
there is a two order-of-magnitude increase in the conductivity
with only a three-fold increase in the concentration. This range is
tentatively identified as the range of the metal–nonmetal transition. Note, however, that there is no discontinuity in the curve.
In the figure one sees that as the temperature is increased the
transition shifts slightly to lower concentrations. If one adopts the
usual expression relating the conductivity σ to the mean free
path, $L = \hbar k_F \sigma / n e^2$, where n is the charge carrier density and k_F

the Fermi wavelength, then the mean free path is found to be 0·02 Å at 2·0 m.p.m. and 3·8 Å at 6·0 m.p.m., while the ion–ion and the dipole–dipole distances are 9·0 Å and 3·6 Å, respectively, at the higher concentration. The simple formula for L is clearly meaningless at 2 m.p.m. and possibly so at 6 m.p.m.

On the other hand, the Hall coefficient (Nasby and Thompson 1970), Fig. 2.6, shows a detectable deviation from free-electron behaviour even at 9 m.p.m. and is about twice the free-electron value at 2·6 m.p.m. yet remains temperature-independent throughout. This concentration, 9 m.p.m., will be taken as the lower limit of the metallic state, and hence the upper terminus of the M–NM transition, in the discussion which follows. The choice is pragmatic and should not be taken as a precise limit (Cohen and Thompson 1968).

The optical data complement the Hall results (Cronenwett and Thompson 1967; Somoano and Thompson 1970). Above 9 m.p.m., the optical data, Fig. 4.2, fit a Drude model, that is, a free-electron model, fairly well, but as the concentration goes

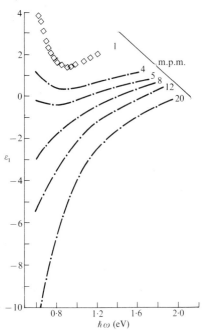

FIG. 4.2. Real part of the optical relative permittivity at the concentrations listed on the right (Thompson 1968).

further below 9 m.p.m. there are changes and the solvated electron line is observed at 1 m.p.m. There is no doubt that bound
electrons contribute to optical properties in the near IR and
visible region, at concentrations of about 5 m.p.m. As an example
there is the data of Beckman and Pitzer (1961). They measured
the reflectivity relative to mercury, already discussed in Chapter
3. At concentrations above 5 m.p.m., they fit the Drüde theory
fairly well when the d.c. values for the electron density and the
relaxation time are used, but below 5 m.p.m., the dropoff in the
reflectivity with wavelength is too rapid, and bound states are
required in their analysis if the data are to be fitted. Note,
however, that they put only a small fraction of the valence
electrons into bound states. The choice of parameters for the
bound states was to some extent arbitrary. Thompson and
Cronenwett (1967) have reported somewhat similar conclusions
from their optical measurements. They measured the real and
imaginary parts of the dielectric constant using a polarimetric
technique as described in Chapter 2. At 8 m.p.m. and above,
their data had the form required by the Drüde theory though
their relaxation time did not fit the d.c. value exactly. The plasma
resonance is clearly discernible in the energy loss function (e.l.f.)
.(Section 2.2.5) above 8 m.p.m. and there is a slight peak in the
e.l.f. even at 4 m.p.m. Electron densities derived therefrom (eqn
2.34) are consistent with Hall effect data (Nasby and Thompson
1970) and with carriers of effective mass near that of the free
electron even at 4 m.p.m. At 5 and 4 m.p.m., trends develop in
the optical data characteristic of bound electrons as may be seen
in Fig. 4.2. Again, however, only a small fraction of the valence
electrons are required to be in bound states. Nevertheless,
Koehler and Lagowski (1969) have shown the optical absorption
near 1 m.p.m. definitely to be identical to that produced by the
solvated electron well below 1 m.p.m.

The most naive interpretation of the Hall coefficient and the
optical data then leads to the conclusion that some of the
electrons must go into bound states below about 9 m.p.m. and in
that sense at least a M–NM transition has begun. Prior to an
elaboration of this tentative conclusion, the data in this concentration region from 1 to 9 m.p.m. are reviewed more systematically.
First those properties showing dramatic changes are presented,
then those which vary less rapidly, if at all, over the range.

The electrical conductivity has already been discussed and a distinct change in the concentration derivative of the conductivity was found at 2 and at 6 m.p.m. In Fig. 2.5, the first logarithmic temperature derivative of the conductivity shows an extremum at approximately 2·6 m.p.m. The second logarithmic derivative is a similar function of concentration. As noted in Fig. 2.6, the Hall coefficient does not show dramatic changes but begins to depart from the free electron value around 9 m.p.m. and shows a maximum departure from the free electron value at about 2·6 m.p.m. Furthermore, the thermoelectric power (Damay, Depoorter, Chieux, and Lepoutre 1970), Fig. 2.8, already discussed extensively in Chapter 2, shows agreement within a factor of two with the simple free electron until the concentration is reduced to the neighbourhood of 3 m.p.m. whereupon it becomes markedly larger than expected on the basis of simple free-electron theory. There is a range wherein the Thomson coefficient $dS/dT < 0$, Fig. 4.3. These transport coefficients then, as a group, show distinct changes in their trends in the neighbourhood of 2·5 and of 6 m.p.m.

The static and spin susceptibility show behaviour quite similar to that of the thermopower. The static magnetic susceptibility agrees with free electron theory within a factor of two until concentrations of about 1 or 2 m.p.m. are reached and then

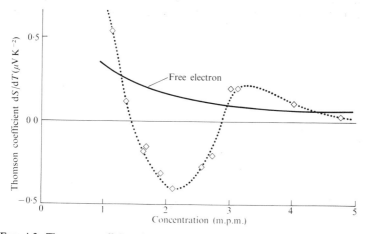

FIG. 4.3. Thomson coefficient (dS/dT). The solid line shows the result expected for free electrons (Damay, *et al.* 1970).

sharply rises as x decreases further (Graper and Naiditch 1969). This sharp rise is observed also when one looks at the spin susceptibility though it is to be noted that there is an extreme paucity of data in each of these cases, Fig. 4.4. O'Reilly (1964a) has reported extensive measurements of the Knight shift at the metal nucleus for sodium solutions and he finds that this Knight shift becomes less temperature-dependent and larger in magnitude at concentrations in the present range, in particular around 1 or 2 m.p.m., Fig. 3.13; though the size of the Knight shift is by no means comparable to the Knight shifts reported for the simple metals in their pure state. Figs 2.14 and 2.15 show values of $\langle|\psi(\text{Na})|^2\rangle$ and $\langle|\psi(\text{Cs})|^2\rangle$ derived from Knight shifts. There are profound changes near 5 m.p.m. (Lelieur and Rigny 1973b). Haynes and Evers (1970) reported similar effects in Li–NH$_3$ solutions near 4 m.p.m. It should be noted that the magnetic relaxation times, as measured in either spin or proton relaxation experiments, show a marked decrease in magnitude and a lowering of the temperature dependence in the range of present concern, Figs 3.9 and 4.5. Here, as in the transport coefficients, there is uniform behaviour among the magnetic parameters yet the concentration range of interest is significantly lower, being near 1 m.p.m.

The mechanical or thermodynamic properties of the solutions also show changes in the concentration range where there appears to be evidence of an M–NM transition. Vapour pressure data

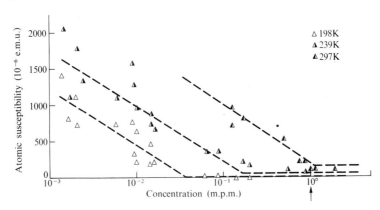

FIG. 4.4. Atomic susceptibility for a number of Na–NH$_3$ solutions (Graper and Naiditch 1969).

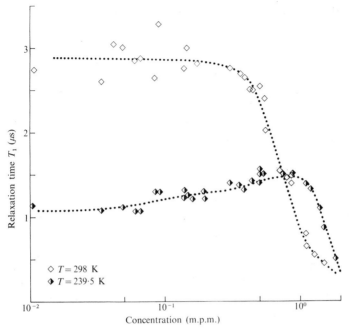

FIG. 4.5. Electron spin–lattice relaxation times T_1 (Pollak 1961).

have been reported for each of the alkali metals dissolved in ammonia by Marshall, as already mentioned in Section 3.5.2. There is a distinct change in the concentration dependence of the vapour pressure, as shown in Fig. 3.23 at concentrations of approximately 6 or 7 m.p.m. Since the vapour pressure data is exceedingly sensitive to the presence of minute amounts of decomposition or chemical reaction, the precise concentration at which the change in the concentration dependence occurs is not thought to be significant because slight errors in the determination of the vapour pressure or of the concentration are likely; however, the fact that there is a distinct change in the concentration dependence of the vapour pressure is thought to be significant here. An integral of the vapour pressure yields the activity, or chemical potential (Dye, Lepoutre, Marshall, and Pajot 1964; Lepoutre 1973; Damay 1973; Ichikawa and Thompson 1973). A computation based on these as well as e.m.f. data is shown in Fig. 3.39. The curves bear remarkable resemblance to those of a pure substance near its critical point (see Chapter 5). Another thermodynamic parameter is the compressibility. Compressibility, either isothermal or adiabatic is difficult to measure directly; however, Bowen and co-workers (Bowen, Parker, and Franceware 1973) have measured the speed of sound in these solutions, which, combined with measurements of the density, may be related to the adiabatic compressibility. Fig. 3.23 shows a rapid rise near 1 m.p.m. Their data may be represented by a linear relationship between the temperature and the speed of sound in the solutions whatever the concentration may be. However, the coefficients in this linear representation show a distinct change in their concentration dependence in the neighbourhood of the metal–nonmetal transition as may be seen in Fig. 3.27. In that figure, the coefficient of the linear temperature term in the solution is plotted as a function of the metal concentration at which the coefficient was determined. In the more dilute solutions the concentration dependence of this coefficient C_1 is weak, being approximately independent of the concentration; whereas at very high concentrations C_1 is smaller and the concentration dependence is again markedly lower. Notice that the range over which the concentration derivative of C_1 shows a discontinuity is approximately 1–10 m.p.m.

Another mechanical property which shows rapid changes in this concentration range is the surface tension, Fig. 3.25, as measured by Sienko (1964). In the case of surface tension, the addition of metal to the solutions produces a decrease in the surface tension; this decrease is small at low concentrations. At the point which has been called the M–NM transition, the *change* in the surface tension with concentration rises rather rapidly and continues to rise with further increases in concentration. Next, it is found that the viscosity, Fig. 3.24, shows a pronounced drop with concentration in the present concentration range. The viscosity only decreases slightly at low concentration but there is a rapid decrease throughout the range in question.

In the main, the data described were taken in the neighbourhood of the normal boiling point of liquid ammonia and therefore do not lend themselves to a determination of the locus of the so-called metal–nonmetal transition in the temperature–composition plane at large. At the moment, the only systematic data which may be directed to the determination of this locus are data due to Catterall (1965) on the asymmetry of the e.s.r. line. Catterall has reported that the e.s.r. line becomes asymmetric at temperatures and compositions which are essentially independent of the solute ion (Chan, Austin, and Paez 1970). The asymmetry of the electron spin resonance line is presumed to result from the increasing conductivity of the solutions and the concomitant decrease in the skin depth of the material; that is, from the decrease in the penetration of the magnetic field used in the electron spin resonance experiment. A theory for this effect has been worked out some time ago by Dyson (1955). As the temperature is increased, the e.s.r. line becomes asymmetric at concentrations which are lower, Fig. 4.6. The precise conductivity at which the spin resonance line becomes asymmetric is not easily determined but by comparing Catterall's data and the electrical conductivity data of Nasby shown in Fig. 4.1, one sees that the e.s.r. line characteristically becomes asymmetric at concentrations significantly below those at which the electrical conductivity becomes large. Nevertheless, these data may be used to determine crudely the locus of the metal–nonmetal transition in the

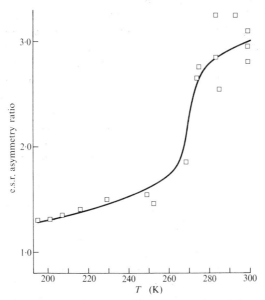

FIG. 4.6. e.s.r. asymmetry ratio at several temperatures for a 1·3 m.p.m. Cs–NH₃ solution. The M–NM transition presumably occurs near 270 K (Catterall 1965).

temperature–composition plane. It appears that the locus of the metal–nonmetal transition is such that increasing temperature produces this transition at lower concentrations though one need not associate the locus of the transition precisely with the locus of the onset of the asymmetry in the e.s.r. line as reported by Catterall.

Those properties of metal–ammonia solutions which show more or less dramatic changes with concentration in the range from 1 to 9 m.p.m.—that is, in the range between concentrations at which the properties are clearly those of an electrolyte and concentrations at which the properties are clearly those of a liquid metal—have been displayed in the preceding paragraphs. Two distinct concentration regions are apparent within the 1 to 9 m.p.m. range: one around 4 m.p.m. where such parameters as the electrical conductivity show dramatic changes and another concentration range around 1 m.p.m. where such properties as the magnetic susceptibility, the Knight shift etc. show dramatic changes. Other points of change may be discerned near 3 and 6 m.p.m.

In contrast to those parameters which show a marked change in the concentration range between 1 and 9 m.p.m., there is a class of properties whose concentration trends do not vary appreciably in this concentration range and any theory proposed for the apparent M–NM transition must account for their relative lack of dependence upon the presumed delocalization of the electron. The properties in this latter category are the following: the density, NH_3 Knight shifts, and positron annihilation data. As may be seen in Fig. 2.18, the excess volume (the volume attributed to the solvated electron in Chapter 3) is essentially independent of concentration over the range of 1 to 9 m.p.m. except that there is a slight maximum in the neighbourhood of 4 m.p.m., at which the volume of the solvated electron is about 10 per cent higher than it is in any other region. These data are exclusively for sodium and potassium and there are inadequate data for the determination of whether or not there is an extremum for any of the other metals which dissolve in liquid ammonia. Next, one finds that the Knight shift (Pitzer 1964) at both the hydrogen and nitrogen nuclei—that is, at the nuclei of the ammonia molecule, are increasing more or less linearly with added metal and that they continue with this trend throughout the concentration range in question.

The temperature coefficients of the Knight shifts $K(H)$ and $K(N)$ at hydrogen and ^{14}N nuclei, respectively, were shown in Fig. 3.12 and the proton data (Hughes 1963) show a distinct drop. The ^{14}N shifts $dK(N)/dT$ measured by O'Reilly (1964) and by Lelieur and Rigny (1973b) do not agree. It is possible that line distortion is responsible for the absence of a drop in O'Reilly's $dK(N)/dT$ at higher concentrations while concentration errors could be the source of large values of $dK(N)/dT$ found by Lelieur and Rigny near 1 m.p.m. The data of Acrivos and Pitzer (1962) scatter too much to resolve the discrepancy. Since both H and N are tied to the same molecule it seems unlikely that $K(N)$ and $K(H)$ could change in different ways with T (Pitzer 1964) and Hughes' data are to be preferred for $dK(H, N)/dT$. This drop is, however, simply a consequence of the drop in susceptibility in this range and there is no evidence that the electron wavefunction near the H or N nuclei is changing rapidly with T, (Hughes 1963; Lelieur and Rigny 1973b). The temperature effects are small so that $K(H, N)$ is still essentially continuous across the concentration range of 9–10 m.p.m.

Finally, and perhaps most surprisingly, the angular correlation of the annihilation gamma rays from positron–electron annihilation in the solutions shows no sign of a concentration dependence in this concentration range. The angular correlation of the γ rays is presumed to reflect the momentum distribution of the electrons in the solution with which the positrons are annihilating (see Section 2.9). These data, which are due to both Varlashkin and Stewart (1966), and McCormack and Millett (1966), show a rather complicated momentum distribution, as may be seen in Fig. 2.17, which is essentially unchanged as one goes from saturation to concentrations of the order of 0·1 m.p.m. and in no sense gives any indication that the electron wavefunction is going from a delocalized to a localized state. The positron lifetime data are similarly insensitive to concentration changes.

Despite the 'well-behaved' nature of these last listed properties in the concentration range between the metallic and the non-metallic state, there is clear evidence that the mobility of the electron, as well as its magnetic interactions, and the free energies of the solutions undergo profound changes in the concentration range between 1 and 9 m.p.m. and in anticipation of the eventual conclusion these changes are described as consequences of a metal–nonmetal transition.

4.3. Data from other systems

4.3.1. Introduction

Before investigating the various mechanisms proposed for the explanation of the M–NM transition in these and other systems, the known properties of three other systems in which metal–nonmetal transitions are said to occur will be examined. The systems may be considered as more or less comparable to the metal–ammonia solutions of paramount interest here. The first system to be considered is that of a metal vapour near its critical point, the second, that of a monovalent impurity in a semiconductor, and the third, a metal–molten salt mixture.

4.3.2. Pure metal vapours

Most of the work on pure metal vapours (Hg, K, Cs) is due to Hensel and co-workers (see Hensel 1973 and references therein) though there have been other contributors (see Even and Jortner 1973a, b and references therein), Parameters measured include conductivity, thermopower, Hall coefficient, optical absorption, and density. The liquid–vapour coexistence curve has also been mapped in part. These are exceptionally difficult experiments due to the reactive materials, high temperatures, and relatively high pressures.

Mercury is divalent and caesium monovalent and therefore differences are to be expected in their behaviour. Figs 4.7 and 4.8 are composites which show these differences. In Fig. 4.7 the electronic parameters are shown for mercury for temperatures slightly above the critical temperature of 1760 K as a function of density ρ. It is significant that most of the changes in both σ and S occur for densities well above that $(5\cdot3 \text{ g cm}^{-3})$ of the critical point. The Hall mobility $(\mu = \sigma R)$ is almost constant at $0\cdot1 \text{ cm}^2 \text{ V}^{-1} \text{ s}^{-1}$ when the density is reduced below 9 g cm^{-3}.

The behaviour of supercritical Cs vapours, as shown schematically† in Fig. 4.8 as functions of density are somewhat different, though there is not so much data as in the case of Hg.

---

† Data for the alkali metals are as yet (Freyland, Pfeifer, and Hensel 1974) not so complete as for Hg. While the critical points are known for K and Cs much of the rest of the phase diagram is schematic only.

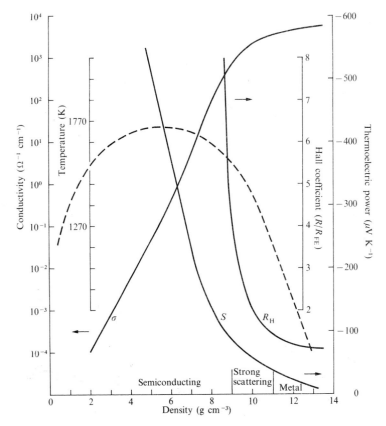

FIG. 4.7. The M–NM transition in Hg vapour as revealed by various data together with the phase-separation curve (dashed). The M–NM transition occurs for densities well above the critical.

Potassium behaves much like Cs. The conductivity at the potassium critical point is three orders-of-magnitude higher than at the equivalent point in Hg.

4.3.3. *Doped semiconductors*

Similarly large changes in conductivity have been observed over narrow concentration ranges in doped semiconductors and are illustrated in Fig. 4.9. Interpretation is complicated by compensation effects (see Mott and Davis 1971, Chapter 6; Allen and Adkins 1972; Mott 1972a).

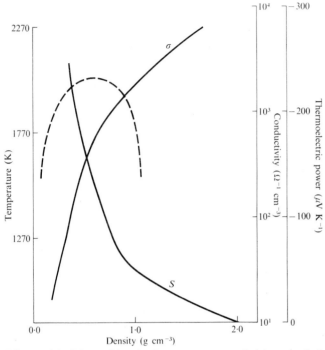

FIG. 4.8. The M–NM transition in Cs vapour as revealed by σ and S data together with the coexistence curve (dashed). The critical density and M–NM transition are coincident.

The resistivity data permit one to separate out three concentration regimes in the doped semiconductors. In n-type silicon for example, at donor impurity levels below about $10^{18}\,\mathrm{cm^{-3}}$, the resistivity behaviour is characteristic of a doped semiconductor. In the region 10^{18}–$3\times10^{18}\,\mathrm{cm^{-3}}$ the activation energy for conduction decreases uniformly (the so-called intermediate region) and there is evidence for variable range hopping ($\sigma \propto \exp(-A/T^{\frac{1}{4}})$, see Mott and Davis (1971)) at low T. Above $\sim 4\times10^{18}\,\mathrm{cm^{-3}}$ the conductivity is essentially metallic, tending to a finite value as $T \to 0$ as for a disordered metal, though up to $10^{19}\,\mathrm{cm^{-3}}$ some decrease of σ with increasing T is shown. For concentrations in the range between 3×10^{18} and $10^{19}\,\mathrm{cm^{-3}}$, the resistivity decreases slowly with increasing temperature. Finally, for concentrations above $10^{19}\,\mathrm{cm^{-3}}$, the resistivity is essentially independent of temperature. The latter data are similar to those commonly

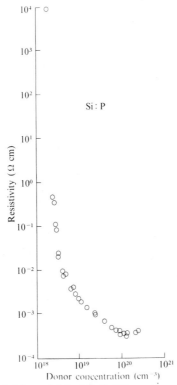

FIG. 4.9. The resistivity at 4·2 K of Si:P. (Alexander and Holcomb 1968).

observed in metals. The effect of compensation is to produce a low temperature conductivity of the form $\sigma \propto \exp(-A/T^{\frac{1}{4}})$ in samples of not-too-low majority concentration.

Some n.m.r. measurements of the metal–nonmetal transition in doped silicon have been made by Sundfors and Holcomb (1964; Alexander and Holcomb 1968; Mott 1974a). They made measurements of the Knight shift and of the relaxation times for phosphorus and boron as well as the silicon nuclei as a function of concentration over the same concentration range as the conductivity data just mentioned. The n.m.r. data show concentration effects clearly in correlation with the three concentration ranges delineated by the resistivity data.

For example, the spin–lattice relaxation time at the [29]Si nucleus in the phosphorus-doped samples showed an inverse square dependence on the donor concentration in the semiconducting state while showing an inverse two-thirds power dependence on the donor concentration in the metallic state with a transitional region corresponding to the transitional region in the resistivity data. Similarly in the metallic state, the product of the spin lattice relaxation time T_1 and the temperature T was a constant as is expected when the nuclear moment interacts with a degenerate distribution of electrons, whereas the temperature dependence of T_1 was weaker in the semiconducting state. Sasaki, Ikehata, and Kobayashi (1973) confirm the drop in ^{29}Si Knight shift below 4×10^{18} donors cm^{-3} reported by Sundfors and Holcomb (1964). It is important to note that Sundfors and Holcomb were able to explain their n.m.r. data only on the assumption that in the boron samples, in the metallic state, the acceptor atoms formed dense clusters within the silicon host; that is to say, there were in-homogeneities with a precipitation of the boron in particular regions of the silicon as if there were two immiscible phases— one, with a high concentration of boron in a metallic state; the other, with a lower concentration of boron and in, perhaps, a nonmetallic state. Jerome (1968) has used a double resonance technique which permits him to look at nuclear resonance in a selected class of nuclei rather than in the bulk material. He found the $T_1 T =$ constant relation to be satisfied near the phosphorus impurities, indicating the electrons to be quasi-metallic there. In contrast the bulk measurements indicate the electrons are localized. The e.s.r. in the doped semiconductors reveals signs of electron delocalization at much lower doping levels than those at which the transport properties become metallic. The single observable line narrows over the range 0·7 to 3×10^8 cm^{-3}, shows a minimum at the upper end, then broadens at higher concentrations due to lifetime effects. The initial narrowing is believed to arise from exchange as the electron moves through larger groups of donors. The paramagnetic susceptibility (Ue and Maekawa 1971) shows no strong variation with doping level though it does increase slightly as the concentration goes down, at 1·5 K.

4.3.4. *Metal—molten salt mixtures*

Solutions of electropositive metals in their molten halides also show wide variations in electronic properties (Raleigh 1963; Bredig 1963; Katz and Rice 1972) but have not received as close an examination in the context of M–NM transitions as have the other systems discussed. Nevertheless there are distinct similarities in phase diagrams (Fig. 5.5) and in other respects as well (Ichikawa and Thompson 1972). Fig. 4.10 is designed to bring out the differences as well as similarities. One sees that while most of the thermopower change etc. has occurred before

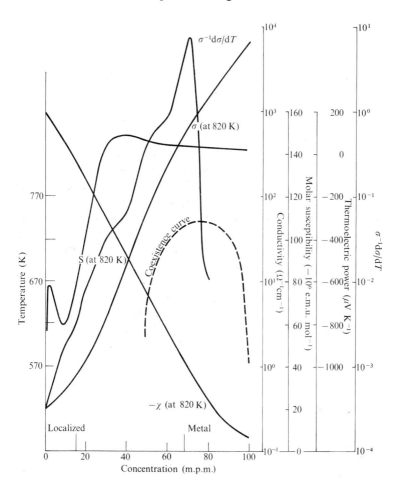

FIG. 4.10. The M–NM transition in Bi–BiI₃ melts as revealed by σ, σ^{-1}, $d\sigma/dT$, χ, and S together with the coexistence curve (dashed). The M–NM transition occurs just to the left of the consolute point (Cubiccioti and Keneshea 1958, 1959; Cubiccioti, Keneshea, and Kelley 1958).

the phase separation the conductivity only attains metallic values at concentrations above the consolute point. The peak molar susceptibility occurs at the concentration of BiI.

4.4. General form of a metal–nonmetal transition

As already noted at the beginning of this chapter there have been a number of surveys of the M–NM transition in a variety of materials and these have been admirably summarized in Mott's book (Mott 1974c).

One can identify a variety of conduction regimes in the interval between metal and nonmetal in noncrystalline materials (Cohen and Jortner 1973b, 1974c; Mott 1973). These are: (a) a truly metallic regime where the mean free path exceeds the interatomic spacing and one can speak of propagation-with-occasional-scattering; (b) a strong scattering, diffusion, or Brownian regime with mean free paths L of the order of the interatomic distance a; (c) an inhomogeneous regime wherein fluctuations produce macroscopic inhomogeneities and concomitant region-to-region differences in conduction mechanism together with boundary effects. Percolation theory is used (Kirkpatrick 1973b). (d) A Mott–Cohen–Fritzsche–Ovshinsky regime where the density of states $N(E_F)$ is not small at the Fermi level E_F, but the states are localized so that conduction is by hopping; and (e) a true semiconducting state wherein there are no states at the Fermi level. Cohen and Jortner (1973b) subdivide region (c) into pseudometallic and pseudoinsulating regions according to whether the volume fraction of the conducting phase is above the percolation threshold (Kirkpatrick 1973b). Mott (1973, 1974c) does not believe that region (c) exists for any random fluctuations of potential on an atomic scale. Cohen and Jortner omit region (d). It is also possible for the localized electrons to produce an ionic or electrolytic regime (as in M–NH$_3$ solutions) to substitute for region (e).

The discussion of a M–NM transition must then include some or all of the processes listed above and not a one-step change. Particularly in sufficiently disordered materials e.g. Si:P, the discontinuous change in free carrier density originally predicted by Mott (1961) is not observed. Mott (1974c) maintains that it is observed in less disordered systems.

4.5. Mechanisms

As there are many stages of a M–NM transition there are many possible origins. Indeed the various mechanisms may even compete for which goes at which density or concentration.

4.5.1. *Correlation*

The classic mechanism is that proposed by Mott (1949) and expanded upon in 1961: the Coulomb interaction of electron and ion requires energy to be supplied for the existence of free carriers—hence at zero absolute temperature the assembly is insulating. That is, for free carriers to be created an electron must be removed from an ion at a cost of the ionization energy I and then donated to a neutral atom with the affinity A as recovered energy. The net excitation energy is $E = I - A$. At $T = 0$ there is no excitation energy available and the system is insulating.

A somewhat similar behaviour is expected in a material with a barely closed gap (semimetal) with few electrons and holes (Mott 1961; Kohn 1964). In such a case the electron–hole interaction will lead to a bound state (an exciton) with, again, an activation energy required for free carriers. A narrow-gap semiconductor is similar (Brinkman and Rice 1973).

In each case it is the long range Coulomb interaction or correlation energy—a fundamentally many-body phenomenon—which is responsible for the existence of the insulating state. Screening reduces the correlation energy and as the interparticle spacing is reduced the screened Coulomb well is no longer able to contain a bound state, the carriers are freed and the material becomes metallic. Alternatively, one may note that the volume over which an electron is localized will decrease as the number density increases. This produces a kinetic energy which must be compared to the Coulomb repulsion. When the density is sufficiently great as to yield a kinetic energy which dominates the Coulomb term the electron is delocalized.

In every case the Coulomb energy must be computed with the dielectric constant ε of the system (e.g. host semiconductor or NH_3) included. Should the dielectric constant be large the Coulomb term is small and the number density n of free carriers on delocalization is decreased. This may be seen as follows.

One calculates the condition that a screened Coulomb well

contain no bound states (Mott and Davis 1971 Section 5.6) to be

$$n^{\frac{1}{3}}a_{\mathrm{H}} \sim 0 \cdot 25 \qquad (4.1)$$

where

$$a_{\mathrm{H}} = h^2 \varepsilon / m^* e^2.$$

Clearly as ε increases n decreases.

Hubbard (1964a, b; Doniach 1970; Mott and Davis 1971) has approached the problem using only single site correlation but has thereby rigorously obtained band splitting. The lower band in such cases corresponds to singly occupied centers (e.g. donors), the upper to doubly occupied sites. The criterion for splitting a half-full band into two 'Hubbard' bands is much like Mott's criterion (eqn 4.1) and one therefore speaks of a 'Mott–Hubbard' transition as being one due to correlation whatever the range and virtually the same formula as eqn (4.1) can be obtained. Hubbard, however, has no discontinuity in n, in contrast to Mott. There is however no contradiction here since Hubbard's treatment does not include long-range Coulomb forces. See also Kittel (1971).

The spins on adjacent sites in the localized state will be oppositely oriented. Indeed, Slater (1951) has pointed out that the onset of antiferromagnetism can lead to an energy gap and a M–NM transition. One thus generally expects antiferromagnetism in the insulating state when Mott–Hubbard localization is involved.

4.5.2. *Polarization*

Herzfeld (1927) and Berggren (1974a, b) have discussed these same ideas from the viewpoint of a 'polarization catastrophe'. The Coulomb interaction is screened by adjacent atoms or molecules rather than by free electrons as in Mott's picture.

4.5.3. *Disorder*

Yet another mechanism for a M–NM transition is due to Anderson (1958, 1970) (see also Ziman (1969), Thouless (1970), and Economou and Cohen (1972)). Anderson showed a regular array of potential wells of random depth produces localization, assuming multi-electron processes may be ignored—a one-electron approximation suffices. Crudely put, it is the inability of an electron trapped on a given site to find another site, in an array of wells of random *depth*, sufficiently close and with a state at the same energy. Without two or more sites with the same levels electron transfer requires activation and the system is nonconducting. Anderson establishes a criterion which states that localization occurs when the ratio of the root mean square deviation of the well depth from the mean well depth, to the bandwidth for an array of wells all at the mean depth, exceeds some specified value. Current estimates put this ratio in the range 2–4 for a cubic lattice (Mott 1974c).

While this calculation has been performed for a regular array of potential wells of random depth, rather than an irregular array of equivalent wells, the general idea that disorder can be a sufficient cause for localization has evolved (Mott and Davis 1971). Of course it is possible for competition between Anderson localization and some other process to occur in a given system.

4.5.4. *Pseudogap*

Mott (see Mott and Davis 1971 and Mott 1974c and references therein) has argued rather persuasively that whenever the density of states $N(E)$ drops too low in a noncrystalline material the electron states at that energy will be localized. Such a low $N(E)$ may occur in a band tail or in the *pseudogap* between two bands. The two cases are sketched in Fig. 4.11. Should E_F lie in this range of energy then nonmetallic behaviour is to be expected. The energies E_v and E_c separate localized and extended

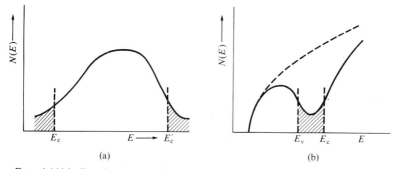

(a)   (b)

FIG. 4.11(a). Density of states in the Anderson model, when states are non-localized in the centre of the band. Localized states are shown shaded. E, E_c' separate the ranges of energy where electrons are localized and non-localized. (b) A weak pseudogap in the density of states; the dashed line is the free electron curve. Localized states are shown shaded.

states and define 'mobility edges' (Cohen 1968, quoted in Mott 1969). The quantitative statement of 'too low' is given in terms of Mott's g-factor (Mott and Davis 1971, Section 2.8)

$$g = N(E_F)/N(E_F)_{\text{free}} \tag{4.2}$$

where $N(E_F)$ is that appropriate to the material in question (e.g. Fig. 4.11b) while $N(E_F)_{\text{free}}$ is that for an assembly of free electrons with an effective mass appropriate to the bottom of the band (see dashed curve in Fig. 4.11b).

The factor g influences many of the properties of the system such as conductivity σ, Hall coefficient R_H electron spin susceptibility χ_{el} and Knight shift $K(N_i)$, and may be determined from these. For intermediate values of g one has

$$\sigma \propto g^2,$$
$$R_H \propto g^{-1},$$
$$\chi_{el} \propto g, \tag{4.3}$$

and

$$K(N_i) \propto g.$$

One thus expects $\sigma \propto R_H^{-2}$, etc. (Even and Jortner 1973a, b; Warren 1971). When $g < 0.3$ it appears that the states near E_F are localized; when $g \gtrsim 1.0$ weak scattering theory (Section 2.4) applies.

4.5.5. Mobility edge and minimum conductivity

The minimum conductivity when E_F is at the mobility edge has been calculated by Mott (1972c) to be near $\sigma_{\text{MIN}} \simeq 0.025\, e^2/\hbar a$ where a is the distance between wells. If a is 3 Å then σ_{MIN} is about $200\ \Omega^{-1}\,\text{cm}^{-1}$. Cohen and coworkers believe that one should take the mobility as going continuously to zero as $E \to E_c$ (say) from above (Cohen and Jortner 1973b).

In the standard texts (Kittel 1971) on solid-state physics, materials are classified as metal, semiconductor, or insulator, according to the Bloch–Wilson scheme which dates from the early 1930s. In this scheme, the description of the material is decided upon by relative positions of the Fermi level and the band edge. If the band structure is such as to provide a completely filled energy band with a gap separating it from the next empty band, then the material will be classified as an insulator. If, on the other hand, there is a band which is nearly filled separated by a gap which is not too large from a band which is slightly filled, then the material will be classified as a semiconductor. Finally, if the band is partially filled, or if bands overlap, then the material will be classified as a metal, as there will be many vacant states in the neighbourhood of the Fermi level.

The effect of pressure on the lattice parameter then might be to open a gap in a divalent metallic material, producing a M–NM transition of the *Wilson* or *band-crossing* type. Mott, however, notes that there is a strong probability of Anderson localization in the pseuaogap whenever the variation of interatomic distance occurs in a divalent liquid e.g. Hg (Mott and Zinamon 1970).

4.5.6. *Inhomogeneity and percolation*

Cohen and co-workers as already stated have recently taken the view that there is an inhomogeneous regime in any M–NM transition (Cohen and Jortner 1973*b*; 1974*a, b, c*; Jortner and Cohen 1974). This section is based on their work. They believe that in many M–NM transitions, particularly those in disordered systems, there will be a stage (Section 4.4) at which the material becomes microscopically inhomogeneous. Some of the material is metallic in nature, containing free electrons in extended states while the remainder is nonmetallic with localized electrons. The various regions are microscopic in size (~40 Å) and are mixed thoroughly. The problem then becomes one of determining whether a free electron can find a (tortuous) path from one metallic region to another and thence across the sample. Percolation theory therefore enters (Kirkpatrick 1973*b*). While results have been worked out for simple systems or for properties depending on gross averages, the problem is immensely complicated, particularly for transport coefficients. This book will only attempt an outline of the work without attempting to resolve the current controversies.

Inhomogeneities arise because of rapid variation of electronic energies, densities and mobilities with the constraints applied to the system. For example, any fluid contains density fluctuations. As a consequence of the sensitivity of, say, mobility to density, changes in fluctuation toward higher density enhances metallic properties in the region of space occupied by the fluctuation while a fluctuation to lower density may lead to localization in that volume. One expects particularly wide excursions from the mean density in a fluid near a critical point (Chapter 5) so that the most obvious circumstance leading to inhomogeneities containing significantly different electronic states is phase separation.

But variations in bonding (Cohen and Jortner 1974b, c) can also lead to a variation in electronic state from point to point. So also can compound formation or clustering in multicomponent systems. Inhomogeneities are therefore more widely found than is phase separation.

One thus visualizes a material much like a vigorously stirred mixture of oil and water but with a microscopic rather than macroscopic scale. In some regions electronic properties are like those of a metal, in the others the electrons are either localized or in almost filled bands.

Drastic simplifications are required for progress. The first is to consider only *two* different sorts of region (oil–water) with different but finite conductivities rather than a continuous spectrum. A semiclassical approach is taken and tunnelling is ruled out. One next gains insight for boundary conditions by considering one phase as a spherical inclusion in the other phase much like the classic problem of a dielectric sphere imbedded in a medium of different dielectric constant (Harnwell 1949).

One is thus led to an effective medium theory: EMT (Kirkpatrick 1973a, b). In such a two-component theory the dominant variable becomes the volume fraction of metallic material C; the other fraction $(1-C)$ is presumed nonmetallic. In the Mott treatment of pseudogaps the mobility depends on electron energy, so in the EMT C is a function of electron energy $C(E)$. The dominant value of $C(E)$ is that for $E = E_F$. With this parameter fixed, one calculates the conductivity by appropriately averaging over the microscopic distribution of metallic and nonmetallic regions of the material.[†] The equations obtained in such a theory are algebraically complicated. Jortner and Cohen in their most recent work (1974) choose instead (when the conductivities are widely different and C small) to use computer simulation of percolation in a cubic lattice to relate conductivity to $C(E)$. If $C(E)$ falls below the critical value for classical percolation C^* then no continuous path through metallic regions exists

---

[†] Other parameters may be treated in an analogous way.

for electrons at that energy. If this condition is satisfied for $E = E_F$, then a M–NM transition occurs. C^* is computed from various models (Kirkpatrick 1973b; Jortner and Cohen 1974) to lie in the range $0 \cdot 15 < C^* < 0 \cdot 30$. As $C(E_F)$ decreases below unity the transport parameters gradually depart from metallic values; there is no abrupt change when $C(E_F)$ reaches C^* though nonmetallic values will be dominant more and more as $C(E_F)$ approaches zero.

In any specific case, say M–NH$_3$ solutions, one must know the course of $C(E_F)$ with concentration, the relation of a given property to C, values of the property of interest in each of the components, and all temperature or density derivatives. Cohen and Jortner (1974c) conclude: 'The numbers extracted from the experimental data should be subjected to close scrutiny for reasonableness, and complementary experimental studies should be initiated to provide a direct test for the existence of the predicted inhomogeneities'. Since M–NH$_3$ solutions provide almost the only system wherein data are sufficient to cover all parameters (Lelieur 1973) further exposition is postponed to Section 4.7 below.

4.5.7. *Electrons in dense vapours*

Electrons injected into dense nonmetallic vapours, such as He, also show a free-to-bound transition at certain vapour densities (Eggarter 1972). The transition is due, in He, to short-range repulsive interactions between the electrons and the gas atoms. These repulsions produce a localized electron state when the fluid density exceeds some critical value. The onset of localization produces an abrupt drop in electron mobility (Levine and Sanders 1962; 1967). Electron localization in polar vapours (e.g. NH$_3$) is also possible as a consequence of attractive interactions of both short and long range (Gaathon and Jortner 1973; McNutt, Kinnison, and Ray 1974).

4.6. Interpretation of data

4.6.1. *Introduction*

We have surveyed briefly a variety of M–NM transitions with names: Anderson, Mott–Hubbard, or Wigner; and without: band-crossing, antiferromagnetism, polarization catastrophe, or inhomogeneous. It is not always possible to know which is which, but salient points of agreement with one model and another will now be given.

4.6.2. *Expanded liquid mercury and caesium*

It is agreed that for densities above $11 \, \text{g cm}^{-3}$ the Hg fluid is metallic and belongs in regime (a) of Section 4.4 (Mott 1972b; Hensel 1973; Even and Jortner 1973; Cohen and Jortner 1974c).

As the density decreases below $11 \, \text{g cm}^{-3}$ the known properties satisfy the relations (4.3) with $g < 1.0$ until g declines to a value near 0.3 at a density near $9 \, \text{g cm}^{-3}$. At densities somewhat lower i.e. $\rho \lesssim 8 \, \text{gm/cm}^3$ there is normal semiconducting behaviour (regime (e) of Section 4.4). The optical energy gap opens up at densities below $5.5 \, \text{g cm}^{-3}$ and reaches the atomic value ($4.9 \, \text{eV}$) at $0.03 \, \text{g cm}^{-3}$ (Hensel 1970). These regions are separated by vertical dashed lines in Fig. 4.7.

It is in the range $7.8–9.2 \, \text{g cm}^{-3}$ that controversy persists. While the high temperature certainly precludes the existence of region (d) of Section 4.4 there is no agreement at present with respect to region (c)—the inhomogeneous regime. Mott (1973) argues that tunnelling eliminates the classical percolation theory. He then (see also Mott and Davis 1971) presumes the localized electrons move along much as do ions in electrolytes (and also Hg^+ or Hg_2^+ ions in the vapour)—certainly without thermal activation. One then cannot expect the simple relations of eqn (4.3) or of semiconductor theory (eqn 3.28) to apply.

Cohen and Jortner (1974b, c) think that it is precisely this range wherein inhomogeneities are important. They believe that the critical metallic fraction C^* ($= 0.2$), for which continuous metallic channels exist throughout the fluid, is at $8.2 \, \text{g cm}^{-3}$. This is the density at which $N(E_F)/N_0(E_F) = C^*$ were the liquid homogeneous, and where $N_0(E_F)$ is the density of states in a macroscopic metallic region. Above $8.2 \, \text{g cm}^{-3}$, the fluid is pseudometallic and EMT equations provide a good fit to R_H, σ and $d\sigma/d\rho$. The average range of density fluctuations is found to be near $15 \, \text{Å}$. C is unity at $9.3 \, \text{g cm}^{-3}$ and zero at $8.0 \, \text{g cm}^{-3}$; $dC/d\rho$ dominates $d\sigma/d\rho$.

There is too little Cs data to provide a real test of the EMT Mott (1974b, c) relates the phase separation to the M–NM transition as does Krumhansl (1965), see Section 5.6. There is a range $1\cdot0 < \rho < 1\cdot2\,\mathrm{g\,cm^{-3}}$ wherein the strong scattering model applies. Below $0\cdot5\,\mathrm{g\,cm^{-3}}$ the semiconducting model applies (Freyland, Pfeifer, and Hensel 1974).

4.6.3. *Extrinsic semiconductors*

These materials are examples of Anderson localization in overlapping Mott–Hubbard bands derived from impurity wavefunctions (Mott 1972a; 1974a, c; Allen and Adkins 1972). Only n-type semiconductors will be discussed: Si:P and Si:Sb. At low concentration (and low temperatures) the donor electrons are localized on donor sites just below the conduction band. Increasing donor concentration results in overlap and in the formation of an impurity band. The electrostatic repulsion of two electrons on the same site splits the donor states so that as the impurity density increases one has two Hubbard bands near the donor energy level. These bands are broadened because of the randomness of positions of donors and because of fields produced by ionized minority impurities (acceptors). At sufficiently high concentrations the two Hubbard bands overlap producing a pseudogap. There may be localized states in the pseudogap and there will be localized states in the tails of the overlapping Hubbard bands. The effect of compensation (i.e. the addition of acceptors) is to lower the Fermi level toward the bottom of the impurity band and toward localized states. One can thus see variable range hopping, regime (*d*) of Section 4.4, at low temperatures.

At low concentrations conduction is by activation from donor levels to the conduction band. At intermediate concentrations carriers are activated from the lower to the upper Hubbard bands with an activation energy which decreases as the bands broaden and overlap. At the highest concentrations the impurity band is metallic and there is no activation.

Mott (1974a, c) has recently discussed the n.m.r. data. Berggren (1973) has shown that the Mott criterion for localization (eqn 4.1) becomes

$$n^{\frac{1}{3}}a_{\mathrm{H}} \sim 0\cdot2$$

where donor wavefunctions appropriate to Si are used. His result is not appreciably different from Mott's and is quite close to the value quoted by Alexander and Holcomb (1968).

4.6.4. *Metal–molten salt mixtures*

As noted these materials have received relatively little attention from the viewpoint of a M–NM transition (Raleigh 1963). Most of the transition is over by the time a 50 m.p.m. mixture is reached, in Bi–BiI$_3$ mixtures at least. Ichikawa and Thompson (1972) find semiconducting behaviour at concentrations just below the consolute point though their results do not fit the constants of eqn (3.28). Katz and Rice (1972) discuss the properties of alkali metal–alkali halide mixtures in the context of Anderson localization etc. They argue that the differences between Na and K mixtures with their respective monohalides are a consequence of the relative positions of Fermi level and mobility edge in the two systems. In the bismuth–bismuth trihalide system there is strong influence from the formation of species such as BiCl or BiBr$_2$. The susceptibility, particularly is influenced by competition between the formation of diamagnetic species such as BiI and formation of paramagnetic species such as BiI$_2$ (Topol and Ransom 1963). Unfortunately the mixtures where such associations are strongest, Bi–BiCl$_3$ and Bi–BiBr$_3$, have not received the attention afforded Bi–BiI$_3$.

The metal–molten salt mixtures are, therefore, not sufficiently well characterized to permit rewarding application of the concepts of Section 4.4 and 4.5. There are nevertheless indications that these mixtures belong in the same class as metal vapours and metal–ammonia solutions.

4.7. Interpretation of metal–ammonia data

4.7.1. *Introduction*

The nonmetallic solutions of alkali metals in liquid ammonia have been shown (Chapter 3) to consist of solvated electrons, solvated metal ions and isolated (non-solvating) NH$_3$ molecules. The localized electron states are therefore apparently associated with well-defined structural units within the solvent (Section 3.10). Though these units depend on the presence of the electron for their stability, they are somewhat similar to the well-defined donor sites in semiconductors and are in contrast to the more diffuse localized states derived from spatial disorder as in metal vapours or molten salts (Katz and Rice 1972). The solvated electrons are spin-paired, most probably in a single cavity about the size of an NH_3 molecule, and probably in association with at least one metal ion. There is on the other hand no convincing proof that the two electrons do not reside in separate cavities. Thus, depending on one's concept of the spin-paired state (Section 3.10.3), there are two rather different low-concentration limits for the metal–nonmetal transition: two electrons in a single cavity or two separate cavities. In the former, Hubbard bands certainly overlap—the medium polarization energy exceeding the correlation energy and the states are akin to Hg. In the latter, wavefunctions must look somewhat like those in an H_2 molecule.

The fact that the sites upon which the electron is localized exist only when the electron occupies the site has not been considered in any of the interpretations offered to date for the M–NM transition. This fact makes the M–NH_3 solutions resemble dense vapours such as He (Section 4.5.7) and the dynamics of electron trapping cannot be ignored. The applicability of the concepts of Section 4.5.1–4.5.6 might be questioned *a priori.* When a metallic solution is diluted the density fluctuations which trap electrons must be isolated and the initial trapping events must produce only singly occupied cavities.

The metallic state, as described in Chapter 2, contains free electrons, solvated metal ions, and isolated NH_3 molecules. It is not clear whether the conduction band is derived solely from the remnants of the cavities or also includes effects from the solvated positive ions.

Much more is known about the properties of M–NH_3 solutions than the materials discussed in the previous sections (or any other materials which exhibit M–NM transitions). Figures comparable to Fig. 4.7, 4.8, and 4.10 could contain much data (Mott 1974*b*); only a sample is shown in Fig. 4.12. A much more detailed picture of the transition should be possible therefore and there have been efforts in that direction since Kraus's time (Kraus 1921; Herzfeld 1927; Bingel 1953; Arnold and Patterson 1964; Lelieur, Lepoutre, and Thompson 1972; Lepoutre 1973; Lelieur, Damay, and Lepoutre 1973). None has been completely successful.

There have been two major attempts to provide a detailed description of this transition which retain credibility at present. One has evolved in the papers of Mott and his co-workers (Mott 1961; 1974*b, c*; Sienko 1964; Catterall and Mott 1969; Acrivos

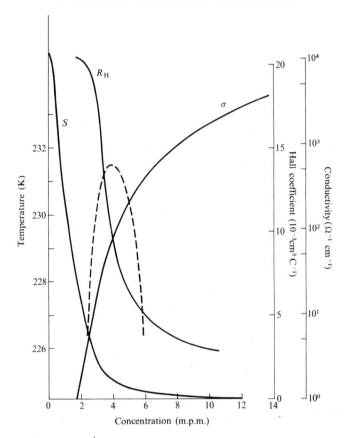

FIG. 4.12. The M–NM transition in Na–NH₃ solutions as revealed by σ, R_H, and S together with the coexistence curve (dashed). The transition occurs just to the left of the consolute point. Compare Figs 4.7, 4.8, and 4.10.

and Mott 1971; Acrivos 1972). It has its basis in the classic Mott transition. Another attempted explanation is primarily due to Cohen (Thompson 1968; Cohen and Thompson 1968; Jortner and Cohen 1973, 1974; Chieux and Lepoutre 1973; Kirkpatrick 1973*a*; Lelieur 1973; Cohen and Jortner 1974*c*; Thompson and Lelieur 1974). It is founded in the notion of percolation in inhomogeneous systems (Scher and Zallen 1970; Kirkpatrick 1973*b*).

These two, somewhat disparate, theories will be discussed separately then compared.

4.7.2. Mott's model

Sienko (1964) and also Kyser and Thompson (1965) pointed out that M–NH$_3$ solutions satisfied Mott's criterion (4.1) for a M–NM transition near 4 m.p.m., when due regard was paid to dielectric screening by NH$_3$.

Mott has provided a detailed exposition of his position quite recently (Mott 1974b, c). He envisions the M–NM transition as being the result of band overlap or band-crossing, with Anderson localization occurring in the pseudogap as described in Section 4.5.4. This conclusion is based on the following observations and assumptions:

(1) the localized electrons are in spin-paired states but in separate cavities; this species is called a 'molecular dimer';

(2) there are two bands: one for an *extra* electron on a molecular dimer and one for a *hole* in a molecular dimer. These are *not* Hubbard bands, rather the bands formed from molecular orbitals (Mott 1974c), the Coulomb repulsion does contribute to the energy gap nevertheless so that the 'Hubbard energy' (Mott 1974c) is important (see also Section 4);

(3) both transitions across the pseudogap and the singlet–triplet transitions within the spin-paired species will have small oscillator strength;

(4) the high-frequency dielectric constant is therefore low and the jump in the number of free carriers n (eqn 4.1) is large;

(5) a discontinuity in n produces a kink in the free energy curve as shown in Fig. 5.10 (Thompson 1967) and a phase separation (Chapter 5).

The M–NM transition occurs upon dilution as the pseudogap opens, with conduction occurring via electrons excited to the mobility edge of the upper band.

Mott separates out two sub-regimes of the M–NM transition: one from 6–3 m.p.m. wherein the bands overlap and the other 3–0·3 m.p.m. in which the material behaves as an intrinsic semiconductor. This model will now be used to describe several of the properties of the solutions.

In the 6–3 m.p.m. range the Mott g-factor should determine σ, χ, $K(N_i)$ and R_H (eqn 4.3). One expects behaviour like that of Hg in the range $9 < \rho < 11 \text{ g cm}^{-3}$ (Section 4.6.2). These data have been treated by Acrivos and Mott (1971) and Acrivos (1972).

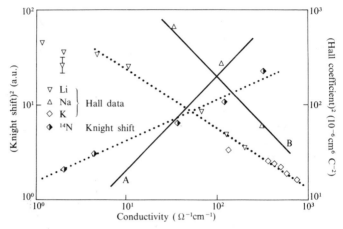

FIG. 4.13. Comparison of M–NH₃ results to Mott pseudogap theory. The expected relationship of Hall data to conductivity is depicted by line B. Knight shifts for ^{14}N should be related to the conductivity by a line such as A. The dotted lines give a 'best' fit to the data. K(Li) show an even weaker dependence on σ (Haynes and Evers 1970).

There are inconsistencies between the values of g deduced from conductivity and magnetic data (Mott 1974c). When g reaches a value of 0·3 near 3 m.p.m., the conductivity is below 50 Ω^{-1} cm^{-1}. This is rather lower than usual estimates of the minimum metallic conductivity.

Fig. 4.13 shows the usual plots of σ *versus* K^2 and σ *versus* R_H^{-2} which should be of unit slope if eqn (4.3) apply. The ^{14}N Knight shifts were measured in Na–NH₃ solutions by Duval, Rigny, and Lepoutre (1968; Lelieur and Rigny 1973b), though there appears to be little solute effect (O'Reilly 1964). The Na–NH₃ conductivities are from Kraus (1921). The Hall data and associated conductivities for K–NH₃ solutions are from Vanderhoff and Thompson (1971), while the Na–NH₃ data are from unpublished work of Suchannek and Naiditch (1966), and the Li–NH₃ data from Nasby and Thompson (1970). Though the three Na–NH₃ points are consistent with $\sigma \propto R_H^{-2}$, the weight of the Li–NH₃ and K–NH₃ data do not fit eqn (4.3) nor do the Knight shifts fit $K^2 \propto \sigma$. One must expect inconsistencies amongst the g-values derived from several sources if eqns (4.3) are not precisely satisfied. Acrivos and Mott (1971) suggest that polaron effects on the conductivity may vitiate eqns (4.3) in M–NH₃ solutions.

The lower concentration range (<3 m.p.m.) is presumed to display semiconducting behaviour. There is a sudden rise in $d\sigma/dT$ (Fig. 3.3) in this range and dS/dT changes sign (Fig. 4.3), both as expected for semiconductors. However, S has a magnitude somewhat less than that derived from eqn (3.28), i.e. 87 μV K^{-1} (Mott and Davis 1971). Also the value of $d \ln \sigma/dS$ exceeds by a factor of two the standard value of $1\cdot16 \times 10^{-2}$ K μV^{-1}, though $\ln \sigma$ is proportional to S (Thompson 1964).

The presence of medium polarization energy (Section 3.10) which does *not* move with an electron from site to site is known (Cutler 1972) to reduce both the thermopower and $d \ln \sigma/dS$ and may well be important in M–NH$_3$ solutions while being a small effect in, say, Si:P. Thus the quantitative differences between the data described here and the expectations for an Anderson transition do not, *a priori*, rule out Mott's interpretation.

Mott ascribes the decrease of σ with pressure (Fig. 2.5) to a contraction of the cavity and a consequent increase in the Coulomb repulsion (Hubbard energy). The effect of pressure on both optical absorption and spin-pairing of the solvated electron has already been discussed in Sections 3.3 and 3.4. There it was shown that the spin-pairing equilibrium (eqn 3.20) was unchanged by pressure and that the absorption maximum shifts by about $2\cdot5 \times 10^{-5}$ eV atm^{-1} (Schindewolf 1968; 1970). Such a pressure shift is consistent with the compressibility of NH$_3$ and requires the cavity to change radius by only about 5 per cent at 1000 atm. Whether such a compression could alter the Hubbard energy sufficiently to produce approximately 50 per cent change in the resistivity is not clear.

The abrupt drop in Knight shift seen when Si:P enters the nonmetallic state would not be expected in liquids such as these were the dipolar interaction the only origin of the shift (Mott 1974*a*). Motional averaging eliminates dipolar terms. However, there is also a shift due to spin contact interaction with the metal nucleus and this is clearly revealed in Fig. 2.14. When the electron is localized at a site remote from the metal nucleus there is a loss of contact interaction and a drop in $\langle|\psi(M)|^2\rangle$.

As with other authors (Krumhansl 1965; Ross 1971) Mott finds the M–NM transition to be related to the phase separation. This point is expanded upon in Section 5.6. The M–NM transition only occurs near 3 m.p.m., i.e. at a concentration below the phase separation. The critical fluctuations, therefore, are not important to the transition, as such. The fluctuations take the system between two *metallic* phases, in this model, and their effect, if any, on the transport properties of the system is small (Section 5.5).

4.7.3. *The percolation model*

Jortner and Cohen (1973; 1974) have presented an effective medium theory (EMT), modified it to include boundary scattering (EMTZ), and applied it to solutions of Li and Na in NH_3, for metallic volume fractions $C(E_F)$ well above C^* ($= 0.17$). They find the EMT of the conductivity σ to be sufficiently different from computer simulations for $C < 0.4$ as to be untrustworthy and use the computer results to provide an interpolation scheme for σ in this range. In the absence of numerical results, they use their EMT or EMTZ only for both Hall coefficient R_H and thermoelectric power S. While the computer fit to σ is entirely satisfactory the treatment of R_H and S is not complete for $C < 0.4$, as expected in the absence of a computer simulation of these parameters.

Their choice of parameters will be given first, then an analysis

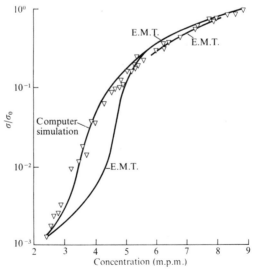

FIG. 4.14. Inhomogeneous model of M–NM transition for σ in M–NH_3 solutions (Jortner and Cohen 1974).

of $d\sigma/dT$ based on EMTZ due to Lelieur (1973; Lelieur and Thompson 1974), followed by a criticism of the EMTZ work and some general comments. The allure of a quantitative theory apparently yielding a variety of parameters will be seen to be somewhat tarnished.

It is unfortunate that the complexity of the problem has prevented Jortner and Cohen from working out a complete, quantitative theory so that their ideas might be tested in full. They choose $C=1$ at 9 m.p.m. and $C=0$ at 2·3 m.p.m. so that the metallic fraction conductivity is $\sigma_0 = 1210\,\Omega^{-1}\,cm^{-1}$ and the nonmetallic is $\sigma_1 = 1·6\,\Omega^{-1}\,cm^{-1}$ for Li–NH$_3$ at 223 K. Their ratio X $(= \sigma_1/\sigma_0)$ is therefore $1·3 \times 10^{-3}$. The C-scale is linear in concentration so that C^* occurs at 3.5 m.p.m. where $\sigma = 18\,\Omega^{-1}\,cm^{-1}$ while $C = 0·4$ at 5 m.p.m.

The choice of a metallic phase at 9 m.p.m. is dictated by the observation of a free-electron Hall coefficient together with a conductivity which is no longer thermally activated above that concentration. Large fluctuations exist only below 9 m.p.m. C is taken as zero at 2·3 m.p.m. as somewhat of a compromise between a linear relation of susceptibility χ to x and a more successful fit of σ to the Cohen–Jortner computer simulation of percolation. Large scale fluctuations persist below 2 m.p.m. in Na–NH$_3$ solutions at 238 K; metallic cluster sizes encompassing 125 ions are found at 2 m.p.m. (Lelieur and Thompson 1974). χ is used to determine the linear relation of C to concentration x. That is, χ is proportional to x (Lelieur and Rigny 1973a) and to C so C is proportional to x. Since the proportionality of χ to x extends above 9 m.p.m., χ cannot be used to fix the extent of the inhomogeneous range.

Fig. 4.14 shows the conductivity as treated by EMT, EMTZ, and computer simulation. Jortner and Cohen do not expect complete agreement for C below 0·4.

In the absence of a better theory EMT is applied to parameters other than σ. Only a brief outline will be given here. The EMT equations for the conductivity are:

$$\sigma = f\sigma_0,$$
$$f = a + (a^2 + \tfrac{1}{2}X)^{\frac{1}{2}},$$
$$a = \tfrac{1}{2}[(\tfrac{3}{2}C - \tfrac{1}{2})(1 - X) + \tfrac{1}{2}X], \qquad (4.4)$$
$$X = \sigma_1/\sigma_0.$$

When boundary scattering is included (EMTZ) eqns (4.4) are replaced by:

$$\sigma = \bar{f}D(C)\sigma_0$$
$$\bar{f} = f[C, X(C)]$$
$$D(C) = ZC/(1 - C + ZC)$$
$$X(C) = \sigma_1/\sigma_0 D(C)$$

(4.5)

where Z is the ratio of the sampling length for concentration fluctuations to the mean free path. Lelieur (1973) has computed $d\sigma/dT$ and $d\sigma/dP$. The pressure and temperature derivatives of C are evaluated from compressibilities and densities of the end point solutions. Lelieur used $C = 0$ at 1 m.p.m. rather than 2·3 m.p.m. He finds the large negative values of $d\sigma/dP$ (Fig. 2.5) to be a consequence of the higher compressibility of the metallic regions when compared to the nonmetallic. It is not clear if his results will endure in a better theory. Jortner and Cohen (1974) computed R_H, S, and the thermal conductivity with success insofar as trends are concerned.

The computations described above all were made for temperatures 10–15 K above the consolute temperature. There is thermodynamic evidence (Lelieur and Thompson 1974) for concentration fluctuations of sufficient *extent* in space at these rather high temperatures. The fluctuation determination is made from the concentration correlation function (Bhatia and Thornton 1970) and is not limited to 'critical' fluctuations. Showing that the fluctuations are long range is not the same as saying that the fluctuation is sufficiently strong in concentration as to give the two limiting concentrations of 2·3 and 9 m.p.m. The spectrum of concentration fluctuations, their spatial extent and their lifetime are not *all* specified by the single measurement of $\langle(\Delta x)^2\rangle$. Jortner and Cohen do not in general limit themselves to critical point fluctuations but in the M–NH$_3$ solutions they only consider the inhomogeneous model applicable over the concentration range of the phase separation. At higher temperatures for Li and Na solutions and in Cs–NH$_3$ solutions they consider the transition to be of the Mott type.

4.7.4. *Comparison*

Unfortunately neither model is entirely successful. There are deviations from predictions derived from Mott's pseudogap factor

g which may be attributable to polaron effects but are nevertheless disquieting. Few parameters have been obtained quantitatively. At the same time the effective medium theory requires substantial modification for success in the $C > 0.5$ range and can only be treated by computer simulation at lower concentrations. The physical origins of the problems with the EMT are recognized but require a more sophisticated theory than has yet been produced. Both sets of authors (among others: Lepoutre (1973)) regard the 3–9 m.p.m. regime as different from that below 3 m.p.m. but above, say, 0.5 m.p.m. Neither provides an adequate description of the nearly two-order-of-magnitude increase in σ in the lower range or of the minimum in χ near 1 m.p.m.

Their shortcomings exceed their differences. Mott admits to critical fluctuation effects over perhaps 5 K above the critical point. Jortner and Cohen require large scale fluctuations over a wide concentration range and for at least 10–15 K above the consolute temperature. However, the main point of disagreement is not so much the nature of M–NH₃ solutions, rather the usefulness of a classical percolation model (Cohen and Jortner 1973a, b; Mott 1973) and the existence of the macroscopic fluctuations required for such a model.

A number of points have not, as yet, been treated. There is the case of e_2^{2-} as the localized electron state. While one might regard this case as analogous to Hg, the absence of a regime where eqns (4.3) are satisfied still requires explanation. Lepoutre (1973) has suggested that the two overlapping bands are derived from solvated electrons (for the lower band) and from solvated cations (for the upper band). There is then no Hubbard energy but the possibility of Anderson localization in the pseudogap remains.

More critical, perhaps, is the careful treatment of the solvated electron state and the dynamics of its formation. Recall that the solvated electron wavefunction penetrates far beyond the first solvation layer. If a solvated electron detects a nearby fluctuation of sufficient depth to produce some slight transition probability, and if the fluctuation has sufficient lifetime, then the electron may *tunnel* over and reorganize the solvent at that point so as to stabilize itself at the new site. It would therefore seem that the conventional Anderson model must be reconsidered so as to allow for the ability of an electron to alter the depth of the wells it encounters (Mott 1974c). While such effects do not seem important at low concentrations where they must also exist, since the conductivity increases with viscosity, they may become significant at higher concentrations.

4.8. Final discussion

The complex transition which occurs as a 9 m.p.m. solution of alkali in NH_3 is diluted to 1 m.p.m. is certainly a M–NM transition but the dominant mechanism is not universally acknowledged at present. Correlation and disorder play a role, so perhaps do fluctuations. Perhaps the situation can be rationalized if not clarified by contemplating the results of the following gedanken experiments.

Suppose a lone electron is placed in NH_3 vapour (Section 4.5.7), what is the lowest density at which it can be trapped? The answer is $\sim 10^{-3}\,\mathrm{g\,cm^{-3}}$. One can be absolutely certain of a localized state at liquid densities.

Suppose a lone Na *atom* is placed in NH_3 vapour. Under what circumstances will it be ionized? In the liquid it is certainly ionized and both cation and electron are solvated (trapped) by NH_3 molecules. The fate of the atom at densities lower than the liquid is yet to be established.

Suppose now these two experiments are repeated in a 1 m.p.m. solution.

The electron now has a number of options. It need not dig its own well but may jointly occupy a pre-existing cavity produced by another electron, *if* doubly occupied cavities are stable. If not, then the added electron must find sufficient NH_3 molecules to form its own cavity. Many, perhaps all, of the NH_3 molecules are parts of the solvation layers of the electrons and cations already present in the solution. These molecules may not be so easily adopted for the solvation of yet another particle. Therefore the added electron may be forced to enter the conduction band. Even if these conflicting options do not exist at 1 m.p.m., they may at 2·3, 3·0, or 4·0 m.p.m. This analysis is further complicated by the embryonic cavities which exist between solvated metal cations which may provide sites for electron trapping.

The detailed fate of an added Na atom is similarly obscure. However, ionization certainly occurs at liquid densities, and the whole system becomes more metallic.

With complexities such as these arising when attention is restricted to single particles the difficulties in producing a theory, which encompasses all the available data at concentrations wherein there are strong interactions amongst all species, become understandable.

REFERENCES

ACRIVOS, J. V. (1972). *Phil. Mag.* **25**, 757.
—— and MOTT, N. F. (1971). *Phil. Mag.* **24**, 19.
—— and PITZER, K. S. (1962). *J. phys. Chem.* **66**, 1693.
ALEXANDER, M. N. and HOLCOMB, D. F. (1968). *Rev. mod. Phys.* **40**, 815.
ALLEN, F. R. and ADKINS, C. J. (1972). *Phil. Mag.* **26**, 1027.
ANDERSON, P. W. (1958). *Phys. Rev.* **109**, 1492.
—— (1970). *Comments Sol. St. Phys.* **1**, 190.
ARNOLD, E. and PATTERSON, A. (1964). *J. chem. Phys.* **41**, 3098.
BECKMAN, T. A. and PITZER, K. S. (1961). *J. phys. Chem.* **65**, 1527.
BERGGREN, K.-F. (1973). *Phil. Mag.* **27**, 1027.
—— (1974*a*). *J. chem. Phys.* **60**, 3399.
—— (1974*b*). *J. chem. Phys.* **61**, 2989.
BHATIA, A. B. and THORNTON, D. E. (1970). *Phys. Rev. B* **2**, 3004.
BINGEL, W. (1953). *Annln. Phys.* **12**, 57–83.
BOWEN, D. E., PARKER, M. H., and FRANCEWARE, L. B. (1973). *J. chem. Phys.* **58**, 2984.
BREDIG, M. A. (1963). In *Molten salt chemistry* (ed. M. Blander), p. 417. Interscience, New York.
BRINKMAN, W. F. and RICE, T. M. (1973). *Phys. Rev. B* **7**, 1508.
CATTERALL, R. (1965). *J. chem. Phys.* **43**, 2262.
—— and MOTT, N. F. (1969). *Adv. Phys.* **18**, 665.
CHAN, S. I., AUSTIN, J. A., and PAEZ, O. A. (1970). In *Metal–ammonia solutions* (eds J. J. Lagowski and M. J. Sienko), p. 425. Butterworths, London.
CHIEUX, P. and LEPOUTRE, G. (1973). In *Electrons in fluids* (eds. J. Jortner and N. R. Kestner), p. 193. Springer-Verlag, Heidelberg.
COHEN, M. H. and JORTNER, J. (1973*a*). *Phys. Rev. Letts.* **30**, 696.
—— —— (1973*b*). *Phys. Rev. Letts.* **30**, 699.
—— —— (1974*a*). *Phys. Rev. A* **10**, 978.
—— —— (1974*b*). In *Amorphous and liquid semiconductors* (eds J. Stuke and W. Brenig), p. 167. Taylor and Francis, London.
—— —— (1974*c*). *J. Phys. (Fr.)* **35**, C4–345.
—— and THOMPSON, J. C. (1968). *Adv. Phys.* **17**, 857.
CUBICCIOTTI, D. and KENESHEA, F. J., JR. (1958). *J. phys. Chem.* **62**, 999.
—— —— (1959). *J. phys. Chem.* **63**, 295.
—— —— and KELLEY, C. M. (1958). *J. phys. Chem.* **62**, 463.
CUTLER, M. (1972). *Phil. Mag.* **25**, 173.
DAMAY, P. (1973). In *Electrons in fluids* (eds J. Jortner and N. R. Kestner), p. 195. Springer-Verlag, Heidelberg.
——, DEPOORTER, M., CHIEUX, P., and LEPOUTRE, G. (1970). In *Metal–ammonia solutions* (eds J. J. Lagowski and M. J. Sienko), p. 233. Butterworths, London.
DONAICH, S. (1970). *Adv. Phys.* **18**, 819.
DUVAL, E., RIGNY, P., and LEPOUTRE, G. (1968). *Chem. Phys. Lett.* **2**, 237.
DYE, J. L., LEPOUTRE, G., MARSHALL, P. R., and PAJOT, P. (1964). In *Metal–ammonia solutions* (eds G. Lepoutre and M. J. Sienko), p. 92. Benjamin, New York.

DYSON, F. J. (1955). *Phys. Rev.* **98**, 349.

ECONOMOU, E. N. and COHEN, M. H. (1972). *Phys. Rev. B* **5**, 2931.

EGGARTER, T. P. (1972). *Phys. Rev. A* **5**, 2496.

EVEN, U. and JORTNER, J. (1973*a*). In *Electrons in fluids* (eds J. Jortner and N. R. Kestner), p. 363. Springer-Verlag, Heidelberg.

—— —— (1973*b*). *Phys. Rev. B* **8**, 2536

FREYLAND, W., PFEIFER, H. P., and HENSEL, F. (1974). In *Amorphous and liquid semiconductors* (eds J. Stuke and W. Brenig), p. 1327. Taylor and Francis, London.

GAATHON, A. and JORTNER, J. (1973). In *Electrons in fluids* (eds J. Jortner and N. R. Kestner), p. 429. Springer-Verlag, Heidelberg.

GRAPER, E. B. and NAIDITCH, S. (1969). *J. chem. Engng. Data* **14**, 417.

HARNWELL, G. P. (1949). *Principles of electricity and electromagnetism.* McGraw-Hill, New York.

HAYNES, R. and EVERS, E. C. (1970). In *Metal–Ammonia solutions* (eds J. J. Lagowski and M. J. Sienko), p. 159. Butterworths, London.

HENSEL, F. (1970). *Phys. Letts.* **31A**, 88.

—— (1973). In *Electrons in fluids* (eds J. Jortner and N. R. Kestner), p. 355. Springer-Verlag, Heidelberg.

HERZFELD, K. F. (1927). *Phys. Rev.* **39**, 701.

HUBBARD, J. (1964*a*). *Proc. R. Soc.* **A276**, 238.

—— (1964*b*). *Proc. R. Soc.* **A277**, 237.

HUGHES, T. R. (1963). *J. chem. Phys.* **38**, 202.

ICHIKAWA, K. and THOMPSON, J. C. (1972). *Phil. Mag.* **26**, 483.

—— —— (1973). *J. chem. Phys.* **59**, 1680.

JEROME, D. (1968). *Rev. mod. Phys.* **40**, 830.

JORTNER, J. and COHEN, M. H. (1973). *J. chem. Phys.* **58**, 5170.

—— —— (1974). Private communication.

KATZ, I. and RICE, S. A. (1972). *J. Am. chem. Soc.* **94**, 4824.

KIRKPATRICK, S. (1973*a*). In *Proceedings Second International Conference on the Properties of Liquid Metals* (ed. S. Takeuchi), p. 351. Taylor and Francis, London.

—— (1973*b*). *Rev. mod. Phys.* **45**, 574.

KITTEL, C. (1971). *Introduction to solid state physics.* Wiley, New York.

KOEHLER, W. H. and LAGOWSKI, J. J. (1969). *J. phys. Chem.* **73**, 2329.

KOHN, W. (1964). *Phys. Rev.* **133**, A171.

KRAUS, C. A. (1921). *J. Am. chem. Soc.* **43**, 749.

KRUMHANSL, J. A. (1965). In *Physics of solids at high pressures* (eds C. T. Tomizuka and R. M. Emrick), p. 425. Academic Press, New York.

KYSER, D. S. and THOMPSON, J. C. (1965). *J. chem. Phys.* **42**, 3910.

LELIEUR, J.-P. (1973). *J. chem. Phys.* **59**, 3510.

——, DAMAY, P., and LEPOUTRE, G. (1973). In *Electrons in fluids* (eds J. Jortner and N. R. Kestner), p. 203. Springer-Verlag, Heidelberg.

——, LEPOUTRE, G., and THOMPSON, J. C. (1972). *Phil. Mag.* **26**, 1205.

—— and RIGNY, P. (1973*a*). *J. chem. Phys.* **59**, 1142.

—— —— (1973*b*). *J. chem. Phys.* **59**, 1148.

—— and THOMPSON, J. C. (1974). *J. Phys. (Fr.)* (Colloque C4) **35**, 371.

LEPOUTRE, G. (1973). In *Electrons in fluids* (eds J. Jortner and N. R. Kestner), p. 181. Springer-Verlag, Heidelberg.

—— and LELIEUR, J.-P. (1970). In *Metal–ammonia solutions* (eds J. J. Lagowski and M. J. Sienko), p. 247. Butterworths, London.

LEVINE, J. L. and SANDERS, T. M. (1962). *Phys. Rev. Letts* **8**, 159.

—— —— (1967). *Phys. Rev.* **154**, 138.

McCormack, K. and Millett, W. E. (1966). *Bull. Am. phys. Soc.* **11**, 201.
McNutt, J. D., Kinnison, W. W., and Ray, A. D. (1974). *J. chem. Phys.* **60**, 4730.
Mott, N. F. (1949). *Proc. phys. Soc.* **A62**, 416.
—— (1961). *Phil. Mag.* **6**, 287.
—— (1969). *Phil. Mag.* **19**, 835.
—— (1972*a*). *Adv. Phys.* **21**, 785.
—— (1972*b*). *Phil. Mag.* **26**, 505.
—— (1972*c*). *Phil. Mag.* **26**, 1015.
—— (1973). *Phys. Rev. Letts* **31**, 466.
—— (1974*a*). *Phil. Mag.* **29**, 59.
—— (1974*b*). *Phil. Mag.* **29**, 613.
—— (1974*c*). *Metal–insulator transitions.* Taylor and Francis, London.
—— and Davis, E. A. (1971). *Electronic processes in non-crystalline materials.* Clarendon Press, Oxford.
—— and Zinamon, Z. (1970). *Rep. Prog. Phys.* **33**, 881.
Nasby, R. D. and Thompson, J. C. (1970). *J. chem. Phys.* **53**, 109.
O'Reilly, D. E. (1964*a*). *J. chem. Phys.* **41**, 3729.
—— (1964*b*). *J. chem. Phys.* **41**, 3736.
Pitzer, K. (1964). In *Metal–ammonia solutions* (eds G. Lepoutre and M. J. Sienko), p. 193. Benjamin, New York.
Pollak, V. L. (1961). *J. chem. Phys.* **34**, 864.
Raleigh, D. O. (1963). *J. chem. Phys.* **38**, 1677.
Ross, R. G. (1971). *Phys. Letts* **34A** 183.
Sasaki, W., Ikehata, S., and Kobayashi, S. (1973). *Phys. Letts A* **42**, 429.
Scher, H. and Zallen, R. (1970). *J. chem. Phys.* **53**, 3759.
Schindewolf, U. (1968). *Angew Chem.* **7**, 190.
—— (1970). In *Metal–ammonia solutions* (eds J. J. Lagowski and M. J. Sienko), p. 199. Butterworths, London.
Sienko, M. J. (1964). In *Metal–ammonia solutions* (eds G. Lepoutre and M. J. Sienko), p. 23. Benjamin, New York.
Slater, J. C. (1951). *Phys. Rev.* **82**, 538.
Somoano, R. B. and Thompson, J. C. (1970). *Phys. Rev. A* **1**, 376.
Suchannek, R. and Naiditch, S. (1966). Final Technical Report, NONR Contract 3437(00). Unified Science Associates Pasadena, California.
Sundfors, R. K. and Holcomb, D. F. (1964). *Phys. Rev.* **136**, A810.
Thompson, J. C. (1964). In *Metal–ammonia solutions* (eds G. Lepoutre and M. J. Sienko), p. 307. Benjamin, New York.
—— (1967). In *Chemistry of non-aqueous solvents* (ed. J. J. Lagowski), Vol. 2 p. 265. Academic Press, New York.
—— (1968). *Rev. mod. Phys.* **40**, 704.
—— and Cronenwett, W. T. (1967). *Adv. Phys.* **16**, 439.
Thouless, D. J. (1970). *J. Phys., C (GB)* **3**, 1599.
Topol, L. E. and Ransom, L. D. (1963). *J. chem. Phys.* **38**, 1663.
Ue, H. and Maekawa, S. (1971). *Phys. Rev. B* **3**, 4232.
Vanderhoff, J. A. and Thompson, J. C. (1971). *J. chem. Phys.* **55**, 105.
Varlashkin, P. G. and Stewart, A. T. (1966). *Phys. Rev.* **148**, 459.
Warren, W. W. (1971). *Phys. Rev. B* **3**, 3708.
Wigner, E. P. (1938). *Trans. Faraday Soc.* **34**, 678.
Ziman, J. M. (1969). *J. Phys., C (GB)* **2**, 1230.

5

LIQUID–LIQUID PHASE
SEPARATION

5.1. Introduction

AMONG the striking properties that have attracted attention to the M–NH$_3$ solutions over the years is the immiscibility of concentrated and dilute solutions. A typical phase separation curve for Na–NH$_3$ solutions (Chieux and Sienko 1970) is shown in Fig. 5.1, and a complete phase diagram was shown in Fig. 1.1. Sienko (1964) has called attention to the fact that these phase separation curves are parabolic in shape (Chieux and Sienko 1970; Teoh, Antoniewicz, and Thompson 1971) in contrast to the cubic shape of most coexistence curves (Fisher 1967, 1969; 1971; Heller 1967; Egelstaff and Ring 1968). In this chapter the characteristics of the phase separation with various solutes will be given, and the customary comparison with other systems and available theory will be made. Critical phenomena in general are the subject of intensive current investigation (Green 1971), and any conclusions regarding the situation in M–NH$_3$ solutions must be tentative only. There is the further complication of the apparent interrelation of the metal–nonmetal transition and the phase separation (Krumhansl 1965; Thompson 1968; Mott 1974 a,b; 1974; Cohen and Jortner 1974) to be assessed.

As shown in Fig. 5.1 a Na–NH$_3$ solution with a concentration near 4 m.p.m. separates into two immiscible fluid phases when the temperature is lowered below 230 K. The more concentrated, less dense, metallic phase floats out on top of the low concentration, electrolytic phase as oil floats on vinegar. The interface is sharp and easily observed because of the contrast between the metallic bronze lustre of the high concentration phase and the inky-blue color of the nonmetallic phase. The consolute point, or critical point, is at C. Differences between the two phases disappear for temperatures above T_C. The first observation of this phenomenon seems to have been made, not surprisingly, by Kraus (1907).

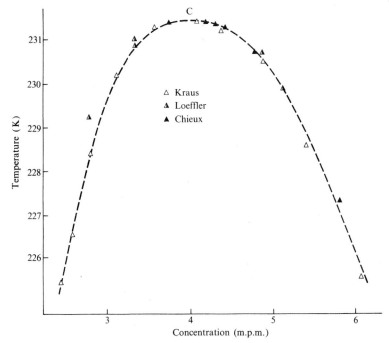

FIG. 5.1. Details of the phase separation in Na–NH₃ solutions (Chieux and Sienko 1970). See the underline for Fig. 2.1 for the notation in this and subsequent figures.

Determination of the locus of the phase separation has generally been made on a visual basis or as a result of vapour pressure changes, but most quantitative studies have been based on the course of electrical resistance with temperature. Figure 5.2 shows typical resistance versus temperature data for sample volumes and concentrations such that the material between the electrodes was primarily metallic (*a*) and equally distributed (*b*) (Teoh, Antoniewicz, and Thompson 1971). When the distribution of the phases is unknown, care must be taken to avoid ascribing more structure to the transition curves than is actually there (Adams 1970; Schürmann and Parks 1971*b*,*c*). The motion of the level separating the two phases, as the temperature is changed can cause apparent anomalies as great as 20 per cent in the resistivity, and the temperature coefficient actually changes by a factor as large as 5 just above T_C. This lends emphasis to the call by

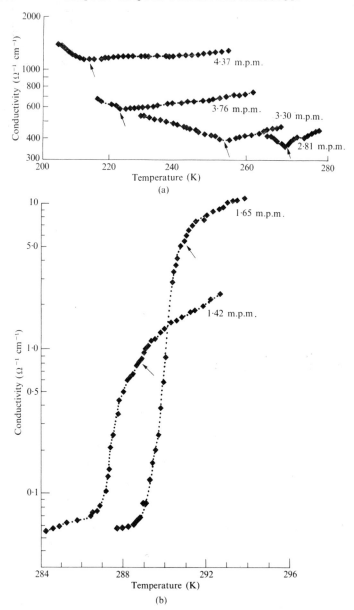

FIG. 5.2. (a) Conductivity as a function of temperature for several Ca–NH₃ solutions. The phase boundaries are marked by arrows. (b) Similar curves for more dilute solutions (Teoh *et al.* 1971).

Sienko (1964) for phase boundary determination by other means. Chieux and Sienko (1970) were unsuccessful in attempts to use differential thermal analysis. D'Abramo, Ricci, and Menzinger (1972) have recently determined a coexistence curve for Ga–Hg using thermal neutron cross-sections (not scattering). Their technique has the added virtue of confirming homogeneity (and the absence of gravitational effects) above T_C. Techniques such as sound speed (Bowen, Thompson, and Millett 1968; Bridges, Ingle, and Bowen 1970) would appear to be potentially useful. Patterson and coworkers have used visual observation in their definitive studies (Schettler and Patterson 1964a, b; Doumaux and Patterson 1967a, b; Schettler, Doumaux, and Patterson 1967). A sealed tube was well stirred then allowed to come to equilibrium in a thermostat. Following visual examination, successively higher and lower temperatures were used until the separation temperature was bracketed within 0.1 K. In some cases the two phases could be sampled for subsequent chemical analysis.

The phase boundaries for Li–, K–, and Ca–NH$_3$ solutions are shown in Fig. 5.3 as determined by Schettler and Patterson (1964a) and by Teoh et al. (1971). There are conflicting reports on the existence of a miscibility gap in Rb–NH$_3$ solutions (Schettler and Patterson 1964a; Sharp, Davis, Vanderhoff, LeMaster, and Thompson 1971) though the most recent work reveals the 'proper' (Schürmann and Parks 1971b) resistance changes. The data are too scanty to require reproduction. There is general agreement that no phase separation occurs in Cs–NH$_3$ solutions, except with the addition of salt (Lepoutre 1970). Separated phases exist in the other soluble metals (Schroeder, Oertel, and Thompson 1969) but the phase boundaries are largely undetermined.

The available data on consolute (critical) points is collected in Table 5.1. Though the solute has little effect upon the consolute concentration, the heavier solutes tend to have lower consolute temperatures insofar as the monovalent metals except for Li are concerned. The divalent calcium solutions depart from both trends: the consolute concentration is lower whether concentration is expressed as m.p.m. or as valence electron fraction (in which case $x_C = 0.033$) and the consolute temperature is much higher. Mixtures of Na and K in NH$_3$ and of Na in ND$_3$ have

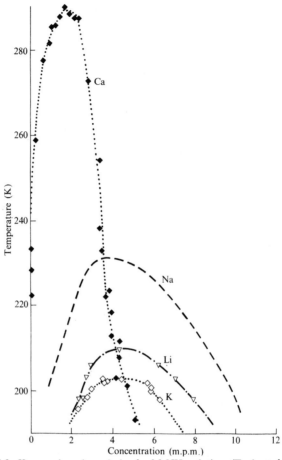

FIG. 5.3. Known phase boundaries for M–NH₃ solutions (Teoh *et al.* 1971).

TABLE 5.1

Critical points of metal–ammonia solutions

Metal ion	Consolute concentration (m.p.m.)	Critical temperature (K)	Critical exponent (β)
Li⁺	4·35	209·7	1/2·1
Na⁺	4·12	231·5	1/2·1
K⁺	4·17	202·9	1/2·0
Cs⁺	*none*	*none*	*none*
Ca²⁺	1·68	290·0	1/2·2

been studied by Schindewolf (Schindewolf, Lang, and Boddeker 1969); the Na–K–NH₃ phase diagram has also been explored by Doumaux and Patterson (1967 *b*). The consolute point in the mixed-metal solutions is close to a linear function of the ratio of, say, mole fraction K to mole fraction Na; whether the departures from linearity are significant or not cannot be resolved at this point. The substitution of ND_3 for NH_3 also appears to produce a linear shift of T_C again without apparent effect on x_C.

The application of pressure, however, does alter the critical composition (Schindewolf, Lang, and Boddeker 1969; Schindewolf 1968). As may be seen in Fig. 5.4, x_C is shifted upward by pressure; the data are sparse but the shift is linear with pressure if one looks at the phase separation temperature at some fixed, *non-critical* composition. An alternate representation of the data can be obtained by using the known isothermal compressibility (Boddeker and Vogelgesang 1971) to replace pressure with volume as a parameter in plotting the critical composition. The

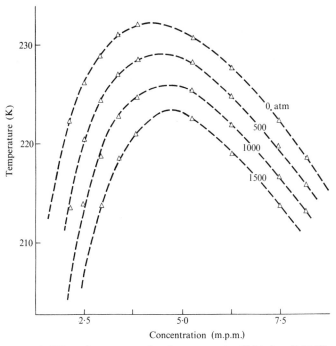

FIG. 5.4. Effect of pressure on the phase separation (Schindewolf 1968).

result is an apparent linearization of the shift of the critical point with pressure. Errors introduced by the necessary extrapolations and interpolations of the meager data weaken the reliability of the result. Nevertheless, it appears that the critical mass density is a linear increasing function of P. Since the mass density and the solute particle density are also linearly related (Naiditch, Paez, and Thompson 1967) in this concentration range, it appears that the critical solute particle density is linearly related to pressure. The critical solute particle density is *not* independent of pressure; this is a point which will appear again.

Solutions with added salt have also been investigated (Cubicciotti 1949; Sienko 1949; Schettler and Patterson 1964b; Doumaux and Patterson 1967a). The added salts are NaBr, NaI, NaN$_3$, and KI, and in each case share a common ion with the metal. The additions all raise the critical (consolute) temperature, broaden the temperature–composition curve and alter the distribution of components. As the total concentration of salt is increased, the dilute phase has increased salt and decreased Na metal concentration; the concentrated phase exhibits the reverse behaviour (Doumaux and Patterson 1967). The magnitude of these effects varies approximately linearly with the amount of added salt as would be expected (Fisher 1968), but depends also on the salt used with the effect increasing with anion size (Damay and Lepoutre 1971).

5.2. Critical phenomena in other systems

Most mixtures and all pure materials undergo some sort of phase separation, and the subject of critical phenomena, in particular, has been widely discussed in recent years. The list of observed critical points is, however, greatly reduced when the concomitant observation of a metal–nonmetal transition is required. Indeed, the nature of the critical point properties is distinctly altered when either or both phases is conducting (Sienko 1964; Egelstaff and Ring 1968; Young and Alder 1971). It is therefore useful briefly to survey such systems as electrolytes, including molten salts, and metal alloys as well as pure metals near their critical points.

Liquid–liquid immiscibility has been observed in common aqueous salt solutions only under conditions of supercooling but is more common in solvents with lower dielectric constants.

Angell and Sare (1968) have interpreted DTA measurements on LiCl–H_2O solutions in terms of phase separation below 160 K and at concentrations near 4 mole per cent salt (the eutectic is near 200 K and 12 mole per cent salt). Lower diffusivity at the low temperatures presumably enhances the phase separation. Hsich, Gammon, Macedo, and Montrose (1972) confirm the immiscibility of aqueous LiCl solutions below the liquidus curve with light scattering data, but find the critical concentration to be close to that of the eutectic. Immiscibility in metal–molten salt mixtures is common (Bredig 1964; Corbett 1964; Sienko 1964) but differs from that already discussed in that in the alkali halides critical compositions are nearer to 50 m.p.m. The fluorides generally have lower consolute concentrations expressed as mole fractions of metal than other metal–metal halide mixtures, though the volume fractions are comparable (Johnson and Bredig 1958). The absence of a phase separation in the heavier metals (e.g. Cs–CsF or Cs–CsI) is comparable to that in Cs–NH_3 solutions, i.e. the consolute point lies below the liquidus as in the aqueous salt solutions. Rb is intermediate between the lighter alkalis and Cs, showing phase separation in all but the Br salts. In salts such as the alkali halides the phase separation and metal–nonmetal transition are quite different from the same phenomena in metal–salt mixtures based on metals which have more than one valence state (e.g. Bi in $BiCl_3$). The phase diagrams are shown in Fig. 5.5. The influence of the extra Bi valence state is clearly visible; a 67 mole per cent metal in $BiCl_3$ solution is equivalent to BiCl.

The most spectacular work on phase separations in conducting fluids is that by Hensel and co-workers on pure Hg, Cs, and K (Hensel and Franck 1968; Hensel 1970; Schmutzler and Hensel 1972; Renkert, Hensel and Franck 1971; Freyland, Pfeifer, and Hensel 1974). See also: Kikoin and Senchenkov (1967), Dillon, Nelson, and Swanson (1966), Young and Alder (1971); Ross (1971); and Even and Jortner (1973). The requisite high pressures and high temperatures have limited the kinds of data available, but there are clear indications that the phase separation is correlated with the M–NM transition in the alkalis but not in Hg, where simple band crossing arguments suffice for the M–NM transition.

The phase separation in Ga–Hg alloys is unaccompanied by a M–NM transition but is nevertheless comparable to that in the

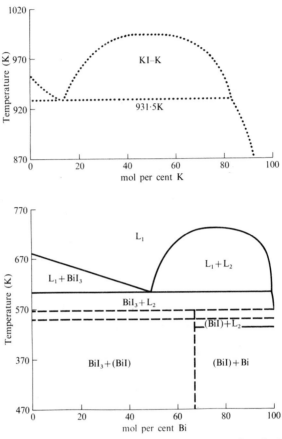

FIG. 5.5. Phase diagrams for metal–molten-salt mixtures (Bredig 1964).

M–NH₃ solutions (Schürmann and Parks 1971a, c; Chieux and Sienko 1970.)

5.3. Critical exponents

The essentials of critical point behaviour are best displayed in the language of critical exponents (Fisher 1965, 1971; Egelstaff and Ring 1968). There is a law of corresponding states (Guggenheim 1945) which includes many non-conducting fluids, and there is uniformity of functional dependence from one material to another. One has, therefore, an economy of description which encompasses many materials. The divergences of many thermodynamic properties at the critical point, for example the isothermal compressibility, may be characterized in terms of the power law which describes the divergence. Other properties, including the shape of the coexistence curve, may also be characterized by a single parameter. For example, if x_1 and x_2 are the concentrations of the two coexisting phases at T, then

$$(x_2 - x_1) \propto (T_C - T)^\beta \qquad (5.1)$$

where T_C is the critical temperature and β is a critical exponent. Nonconducting systems have β near $\frac{1}{3}$, while conducting systems, such as those discussed in this chapter, exhibit β of $\frac{1}{2}$ if

$$\varepsilon \equiv \frac{|T_C - T|}{T_C} \gtrsim 10^{-2} \qquad (5.2)$$

As ε approaches zero β changes over to $\frac{1}{3}$ (Chieux and Sienko 1970). Other critical exponents γ, α, δ may be defined for the following properties of a fluid.

$$K_T^{-1} \sim |T - T_C|^\gamma \qquad (5.3)$$

$$C_v^{-1} \sim |T - T_C|^\alpha \qquad (5.4)$$

$$|P - P_C| \sim |\rho - \rho_C|^\delta, \quad T = T_C. \qquad (5.5)$$

Strictly, the exponent must be defined by a limiting process such as:

$$\beta = \lim_{\varepsilon \to 0} \frac{\ln (x_2 - x_1)}{\ln \varepsilon}$$

The quantity $x_2 - x_1$ is the order parameter for this process (Kadanoff et al. 1967).

For a binary mixture, such as is of current concern, the pressure is replaced by chemical potential (or activity) and density is replaced by concentration. Thus, eqn (5.5) goes over to

$$|\mu - \mu_C| \sim |x - x_C|^\delta. \qquad (5.6)$$

Table 5.2a defines the exponents for a binary mixture (Ichikawa and Thompson 1973). Table 5.2b collects the available critical exponents for conducting and nonconducting fluids (Egelstaff and Ring 1968) at temperatures not too close to T_C. The basis for some of the parameters in the Table is the assumption that there are homogeneity or scaling relations (Fisher 1971)

TABLE 5.2a

Critical exponents for a binary fluid mixture[a]

Exponent	Definition	Measured parameters
α	$\bar{C}_{P,x}$ or $C_{P,x} \sim \varepsilon^{-\alpha}$	Partial or mean specific heat at constant pressure and concentration.
	$\kappa_{T,x} \sim \varepsilon^{-\alpha}$	Isothermal compressibility at constant concentration.
β	$(x'' - x')_{T,P} \sim (-\varepsilon)^{\beta}$	x'', x' are concentrations on each side of coexistence curve.
γ	$\bar{C}_{P,\Delta}$ or $C_{P,\Delta} \sim \varepsilon^{-\gamma}$	Partial or mean specific heat at constant pressure and constant $\Delta (= \mu_1 - \mu_2)$.
	$\kappa_{T,\Delta} \sim \varepsilon^{-\gamma}$	Isothermal compressibility at constant Δ.
	$(\partial^2 G / \partial x^2)_{T,P} \sim \varepsilon^{\gamma}$	Isothermal, isobaric "stability."
δ	$|\mu_i - \mu_{ic}|_{T,P} \sim |x_1 - x_{1c}|^{\delta}$	Ref. b.
$\beta\delta$	$\mu_i - \mu_{ic} \sim \varepsilon^{\beta\delta}$ or $\Delta' \sim \varepsilon^{\beta\delta}$	Ref. b, c.

[a] See Stanley (1971). [b] Constant concentration measurements must be at the critical concentration x_c and constant temperature measurements must be at the critical temperature T_c. The reduced temperature ε is defined by $\varepsilon = (T - T_c)/T_c$. [c] Here $\Delta' = (\mu_1 - \mu_{1c}) - (\mu_2 - \mu_{2c})$.

TABLE 5.2b

System	α	β	γ	Range of ε	δ
Na–NH$_3$	0·2[a,b]	0·502[c]	1·1	$\varepsilon > 10^{-2}$	2·7
	0·7[a,b]	0·34[c]	0·6	$10^{-4} < \varepsilon < 10^{-2}$	
Ga–Bi	0·2[a,b]	0·4[b,d]	1·0[a,b]	$\varepsilon > 10^{-2}$	3·6[b,d]
			1·0[l]		
Ga–Hg	0·3[e]	0·335[e]	1·03[a,e]	$10^{-4} < \varepsilon < 10^{-2}$	4·07[a,e]
		0·373[f]		$10^{-4} < \varepsilon < 10^{-2}$	
		0·5[g]		$10^{-3} < \varepsilon < 10^{-2}$	
Al–In		0·5[g]		$10^{-3} < \varepsilon < 10^{-2}$	3·0[b,h]
CCl$_4$–C$_7$F$_{14}$	1·0[a,b]	0·335[i]	1·2[i]	$10^{-4} < \varepsilon < 10^{-3}$	4[i]
		0·333[j]		$10^{-5} < \varepsilon < 10^{-4}$	
			1·15[k]	$10^{-4} < \varepsilon < 10^{-2}$	

[a] Indicates value calculated from scaling equations. [b] Values due to authors cited. [c] Chieux and Sienko (1970). [d] Predel (1960). [e] Schurmann and Parks (1971a, 1972). [f] D'Abramo, Picci, and Menzinger (1972). [g] Egelstaff and Ring (1968). [h] Yazawa and Lee (1970). [i] Heller (1967). [j] Thompson and Rice (1964). [k] Chu, Thiel, Tscharnuter, and Fenby (1972). [l] Wignall and Egelstaff (1968).

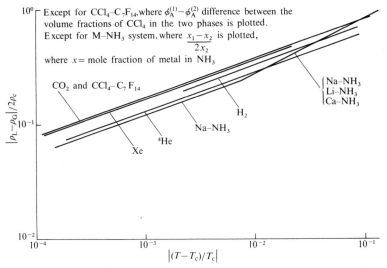

FIG. 5.6. Logarithmic plots of composition versus temperature to show relation of critical exponent β between M–NH$_3$ solutions and other systems. All the dependences are describable by $\beta = \frac{1}{3}$ except for the M–NH$_3$ at the right, for which $\beta = \frac{1}{2}$ (Chieux and Sienko 1970).

among the exponents such as

$$\alpha + 2\beta + \gamma = 2 \tag{5.7}$$

and

$$\alpha + \beta(\delta + 1) = 2. \tag{5.8}$$

There is clear need for further experimentation, especially since there may well be more than one law of corresponding states for conducting materials (Young and Alder 1971). Figure 5.6 shows the crossover from $\beta = \frac{1}{2}$ to $\beta = \frac{1}{3}$ for Na–NH$_3$ solutions (Chieux and Sienko 1970) along with some other data. It must therefore be that each of the exponents of Table 5.2 at some point changes for the conducting systems.

One of the presumed advantages of the critical exponent approach is the presumed universality of the relations. The peculiarities of specific interactions (Coniglio 1972a, b; Swift 1973) are washed out by the long range fluctuations near the critical point. Despite the flaws which are appearing in this concept (Fisher 1971), there is much to be gained by retaining the critical exponents as at least a starting point. Fisher (1968) has developed a renormalization of the exponents which is designed to account for constraints or hidden variables which may not immediately be apparent. For example, if the system contains impurities, the observed exponents will be shifted. The renormalized exponents are of the form $\beta \to \beta/(1 - \alpha')$, where α' is the heat capacity exponent for $T < T_C$ and is approximately the same as α (defined for $T > T_C$). Since α' is near zero in conducting fluids (Egelstaff and Ring 1968) the shift will be small. Fisher suggests that the shifted behaviour will only occur when ε reaches some crossover temperature dependent on, say, the concentration of impurities and on α. A mobile electron gas may also serve as a source of hidden variables and cause a shift. The value of ε at crossover is proportional to the α^{-1} power of the impurity concentration—which in the case of an electron gas might be quite large. Perhaps it is this mechanism which is responsible for the crossover in M–NH$_3$ solutions.

Ichikawa and Thompson (1973) reported an extensive investigation of critical exponents derived from their chemical potential determinations (see also Chapters 2 and 3). As noted in Table 5.2a one can obtain estimates of γ, δ and the product $\beta\delta$. The experiments were flawed by too large values of ε and by possible gravitational effects but the exponents quoted in Table 5.2b were reported (note that some values came from the application of eqns 5.7 and 5.8). For $\varepsilon > 10^{-2}$ the results are consistent with mean field theory. As is required by nonclassical theory, the exponents at smaller ε tend toward the normal values. More precise data would be welcome.

5.4. Multicomponent mixtures

A separate consequence of the presence of impurities is a shift of the critical temperature itself (Fisher 1968). The shift is expected to be linear in the impurity concentration—in agreement with the observations of ND$_3$–NH$_3$ and also K–Na mixtures. The absence of higher order correction terms in the data is surprising. The presence of dissolved decomposition products such as amide ion NH$_2^-$ (see Chapter 1) could also be expected to shift T_C from the value characteristic of the true binary solution. One can speculate that the differences in critical parameters (T_C and especially x_C) among the alkalis may well be due to the differing solubilities of the amides. In any case, control of cleanliness is as necessary here as elsewhere. Finally, if ε is held constant, the variation of the order parameter $(x_1 - x_2)$ will follow the same power law in the impurity concentration as in the temperature. That is, eqn (5.1) is replaced by

$$x_2 - x_1 \propto |x^K - x_0^K|^\beta, \tag{5.9}$$

or

$$|x_2 - x_1|^{1/\beta} \propto x^K,$$

where x^K is the total potassium impurity concentration in a Na–NH$_3$ solution. One has, of course, a three-component system and the point x_0^K is the point in the composition diagram equivalent to the critical point in the P–V plane. It is called the plait point (Clark 1968; Clark and Neece 1968; Griffiths and Wheeler 1970). The plait point is a function of temperature, and the properties of the system near the plait point should vary much as near the critical point. The data of Doumaux and Patterson (1967b), Fig. 5.7(a), may be analyzed in this fashion to obtain Fig. 5.7(b). Clearly $\beta = \frac{1}{2}$ is indicated; there is no sign of a crossover. There is no plait point at 198 K as the system is below T_C for each pure binary; nevertheless, eqn (5.9) applies and β may be determined. The choice of order parameter is somewhat arbitrary. In this system the simple form of eqn (5.9) is obtained only if the LHS is taken as the difference of total metal (Na + K) between the two phases. Model calculations (Clark 1968) indicate a somewhat different choice.

Solutions of salts in M–NH$_3$ solutions should also affect the critical behaviour. The perturbation is more severe than the substitution of K for Na and the results more complex (Cubicciotti 1949; Sienko 1949; Schettler and Patterson 1964b; Doumaux and Patterson 1967a). An analysis of the sort shown in Fig. 5.7 may also be made. The best data are for NaI in Na–NH$_3$ at $-33\,°C$. This temperature is above the consolute point for pure Na–NH$_3$, so that a minimum of salt is required for separation and there is no intercept on the Na-only axis. Figure 5.8(a),(b) shows these results (Doumaux and Patterson 1967a; Schettler and Patterson 1964b). Data for Na–NaBr–NH$_3$ solutions (Doumaux and Patterson 1967a) may also be fit. Clear confirmation of $\beta = \frac{1}{2}$ is again obtained together with indication of the essential similarity of salt and metal solutions. No account of the *ion* concentration is required. The choice of order parameter is this time in

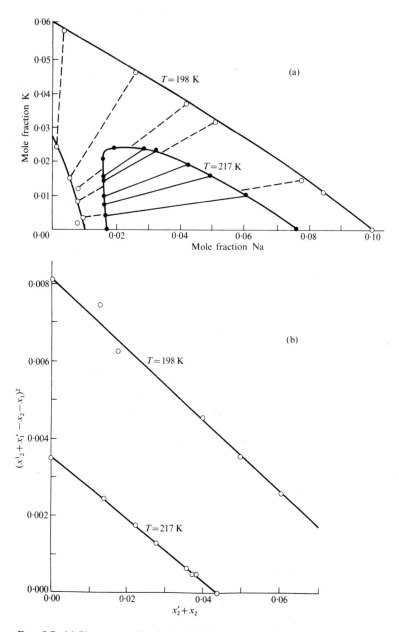

FIG. 5.7. (a) Phase separation in Na–K–NH$_3$ solutions. The phases which coexist are connected by solid tie lines at 217 K and dashed tie lines at 198 K. There is no plait point at the lower temperature (Doumaux and Patterson 1967b). (b) Analysis of the data of Fig. 5.7(a) in terms of eqn (5.9) with $\beta = \frac{1}{2}$.

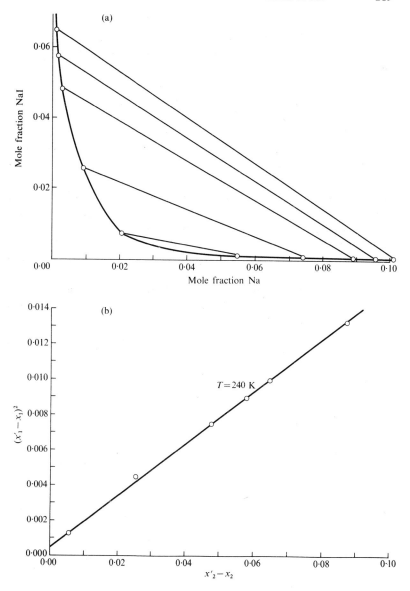

FIG. 5.8. (a) Phase separation in Na–NaI–NH₃ solutions at 240 K (Doumaux and Patterson 1967a). There is no phase separation unless *both* salt and metal are added to NH₃. (b) Analysis of the data of Fig. 5.8(a) in terms of eqn (5.9) with $\beta = \frac{1}{2}$.

exact accord with Clark's (1968) prediction i.e. the difference in sodium metal between the two phases is plotted *versus* the difference in salt. The NH₃ fraction is essentially constant and may be ignored. The curve of Fig. 5.8(a) is almost exactly that predicted by Neece (1967) and by Clark.

Experimental work on another three-component system has been reported by Zollweg (1971). As the temperature is lowered, phase separation occurs without salt, as before. The diagram of Fig. 5.8(a) then changes in a complex way, as sketched schematically in Fig. 5.9. The hash marks indicate the two-phase side of each phase boundary. The original data are not shown (Schettler and Patterson 1964*b*) because some of the tie-lines connecting high and low concentration data pairs cross as a consequence of experimental error, probably because of some water remaining in the salt added to the metal solution. There is clear need for more data as usual. The mixed phase region expands in the usual way as *T* is lowered, when the metal concentration is low. At high metal concentration one apparently finds a two-phase region converting to one phase, then back to two, as *T* is decreased at a point such as A. At B, a lowering of *T* converts the system from one to two phases. It is tempting to conclude that the data at

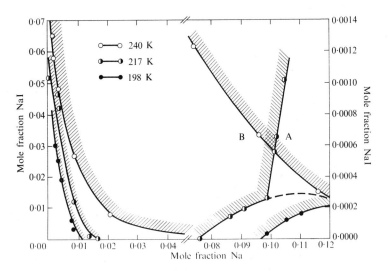

FIG. 5.9. Na–NaI–NH₃ phase separation at several temperatures (Schettler and Patterson 1964*b*). Note the break in the horizontal scale.

217 K are in error near point A. Neece (1967) has constructed a more reasonable set of bimodal curves for a model system which would suggest that the 217 K curve should lie entirely between the other two as shown dashed in the figure.

Each example above is derived from a ternary system, and it is amusing to sketch schematically the ternary phase diagrams so that plait points may be indicated. Since there are known phase separations in the metal–molten salt systems, extreme supercooling may be assumed for the metal–salt binaries so that the existence of separated phases along that leg might be included. It would be consistent with Angell and Sare (1968) if separated phases were also assumed to exist along the NH_3-salt leg of the diagram. It would be interesting to look at the effect of added NH_3 to the metal–molten salt mixtures to see if a lowering of the critical point is indeed obtained. Studies of other ternaries should lead to ready determinations of β in such systems.

Widom (1973) has called attention to the existence of higher-order critical points (e.g. tricritical points) in multicomponent fluid mixtures. Such points arise from the confluence of conventional critical points in systems of three or more components. While none are known in M–NH_3 solutions the possibility exists when there is more than one metal or more than one solvent such as a methylamine–ammonia mixture.

5.5. Resistivity near T_C

In addition to the static aspects of critical phenomena there are dynamic effects as well. In simple systems such as Ar the favourite is light-scattering (critical opalescence) which couples directly to the fluctuations. Schurmann and Parks (1971c) have observed critical point effects on the resistivity and its temperature derivative in Li–Na binary mixtures. They find the temperature derivative of the resistivity to scale as $dR/dT = a + b\varepsilon^{-\lambda}$ where $\lambda = 0.59 \pm 0.05$ at constant composition and a and b are constants. At constant $T = T_C$, the temperature coefficient scales as

$$dR/dT = C + d\,|x_C - x|^{-\mu},$$

where $\mu = 0.7 \pm 0.1$ and C and d are constants. Bends in the M–NH_3 data similar to those analyzed by Schurmann and Parks can be seen in Fig. 5.2 and a similar interpretation is to be expected. The connection to critical fluctuations is presumably made through the dynamic structure factor $a(k, \omega)$ (see Chapter 2) or, in other notation $S(k, \omega)$ (Egelstaff and Ring 1968; Egelstaff and Wignall 1970).

An analysis of the resistivity using the Faber–Ziman approach has been made by Deutsch (1972) with hydrodynamic treatment of the structure factor. See also Lekner and Bishop (1973) for an analysis in the language of Chapter 4.

Similar anomalies have been reported in critical mixtures of isobutyric acid and water (Stein and Allen 1971, 1973; Gammell and Angell 1972). There the effect seems to be an artefact derived from gravitational forces (Jasnow, Goldburg, and Semura 1974). Whether or not a similar explanation can apply to the metal mixtures is yet to be established but is certainly possible (Mott 1974a,b).

Fisher's (1968) renormalization work indicates that $\lambda = \mu(1-\alpha)^{-1}$, which is certainly consistent with Table 5.2 and the errors quoted by Schürmann and Parks. Crossover effects have not been reported.

The nature and origin of the crossover is yet to be fully understood (Fisher 1968). Indeed the origin of the $\beta = \frac{1}{2}$ behaviour is also unclear. Classical theory, such as the simple van der Waals model of a fluid, leads to $\beta = \frac{1}{2}$. More modern, though necessarily approximate calculations support the $\beta = \frac{1}{3}$ found in nonconducting fluids and in conducting fluids for $\varepsilon \leqslant 10^{-2}$ (Fisher 1968, 1971). The usual argument is that classical, or mean-field, values of the critical exponents obtain when the interactions are of very long range (Widom 1962; Coniglio 1972a,b; Swift 1973). Fisher (1968) has pointed out that impurities can also cause a crossover, as in renormalization. Hohenberg (1968) supposes that the crossover must occur when the range of the interaction responsible for the condensation exceeds the range of fluctuations (Brout 1965; Kadanoff et al. 1967). Chieux and Sienko (1970) estimate that there are more than 2000 ions, and/or solvated electrons involved in a single fluctuation at the Na–NH$_3$ crossover point. The long-range Coulomb interaction among the positive ions of a metal is generally given as the basis for the classical behaviour. However, it would seem that screening by the electron gas would be sufficient to reduce the range to a magnitude comparable, say, to van der Waals. Antoniewicz (1970) has suggested that the long-ranged electron wavefunction may carry the effect of the ion–ion interaction to great distances and thus restore the conditions for mean field theory. The current information is clearly limited and requires considerable extension before conclusive statements can be made.

5.6. The metal–nonmetal transition

It has been suggested by Mott (1961, 1974b) in his major work on the transition to the metallic state that the phase separation described here is associated with the metal–nonmetal (M–NM) transition. Krumhansl (1965) has also attributed the phase transition in M–NH₃ solutions to the M–NM transition. Pitzer (1958) had earlier argued that the phase separation in the M–NH₃ solutions was the analogue within a liquid ammonia medium of the liquid metal–metal vapour separation in the pure metal. Inasmuch as this suggestion preceded most of the insight provided by the concept of critical exponents, and was also in advance of the clarification of the nature of the dilute M–NH₃ solutions, the closeness of his suggestion to the current ideas is remarkable. It is clear from the previous discussion that there is rapid variation of the thermodynamic properties near the M–NM transition. In other words, there is a sufficient change of free energy associated with the change of electronic configuration involved in passing from the metallic to nonmetallic state to influence significantly the thermodynamic properties of the entire system. The supposition of Mott (1961, 1974a) that it is primarily a change of electron configuration which is responsible for the M–NM transition and which furnishes the change in free energy driving the phase separation is consistent with these facts (Cohen and Thompson 1968). One may thus postulate free energies appropriate to metallic cohesion at high concentration, or small spacing between the metal ions, and a free energy more consonant with Einstein oscillators or a van der Waals fluid at high spacing. The electronic contributions to the free energy are clearly different in the two extremes. A fairly accurate description of the electronic contributions to compressibilities has been given for densities corresponding to the melting point or below (Ashcroft and Langreth 1967). The end result is a free energy diagram similar to that of Fig. 5.10. As the temperature is

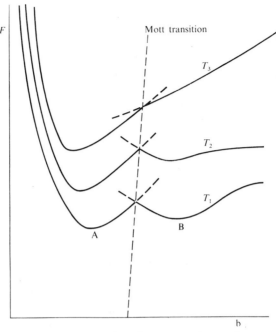

FIG. 5.10. A schematic diagram of the free energy of a metal–ammonia solution in the temperature range of the miscibility gap. The particle spacing is designated b (Thompson 1967), $T_1 < T_2 < T_3$.

lowered one expects a phase separation at the concentrations defined by the points of common tangency. (The curves drawn here are, of course, schematic only.) The minima at A and B correspond to stable configurations in the metallic and insulating phases, respectively. The net result would be a phase separation such as in Fig. 5.1 with the M–NM transition at concentrations tending to x_C as T approaches T_C from above.

A number of possible mechanisms for the metal–nonmetal transition were discussed in the preceding chapter, and any one might also be responsible for the phase separation (Krumhansl 1965). The identification of $\beta = \frac{1}{2}$ suggests that any model providing two adjustable parameters for the constants of the van der Waals equation of state will provide an 'explanation' of at least the shape of the curve. Significant models must also provide an understanding of the trends of the critical point with solute, mixtures of solutes, and mixtures of solvent as well as pressure.

The model of Schettler, Doumaux, and Patterson (1967) suffices for the former but not the latter. They have omitted any electronic contributions to the free energy. It appears that the approach used on metals by Young and Alder (1971) may also apply in the present case. They go one step beyond the van der Waals model by using mean field corrections from perturbation of the hard sphere results. The hard sphere diameter is fixed by fitting the observed structure factor $a(k)$ (see Chapter 2) to the Percus–Yevick formula. This parameter is then combined with the density to obtain a packing fraction, which then permits the pressure (in mean field approximation) to be calculated as a function of density and temperature. The critical parameters are then calculated in the usual way from the equation of state. The results are semiquantitative only, with the critical temperature showing closest agreement with expected values. [Note that T_C has been generally obtained by estimation rather than experiment. Density *versus* temperature data are extrapolated using the law of rectilinear diameters to obtain critical parameters. See, for example, Dillon, Nelson, and Swanson (1966) or a number of papers by A. V. Grosse.] They find T_C directly proportional to the molar cohesive energy and molar volume (Gschneidner 1964) and inversely proportional to the hard core radius (Ashcroft and Lekner 1966). The increase in the hard core with increasing atomic number dominates the increase of molar volume–cohesive energy product in the alkalis, so that T_C decreases in going from Li to Cs as is observed while the opposite trend is predicted and observed in the alkaline earths. This approach, despite its shortcomings, shows ways in which generally available data can be used to improve upon the simple van der Waals model.

An application of these ideas to M–NH₃ solutions has not been made, though the decrease of T_C with increasing atomic size is consistent with Table 5.1. The role of the metal–nonmetal transition is not considered by Young and Alder though they find significant differences between the alkalis and Hg. See also Stroud (1974). Hensel (1973) has concluded that the M–NM transition in Hg is at a density different from the critical density in contrast to Cs where the two are close.

There is no way to ascertain on the basis of existing data whether the M–NM transition is caused primarily by the phase change or the latter takes place because of the free energy change associated with the M–NM transition. Krumhansl finds the M–NM transition 'conducive' to the phase separation, but there are alternate reasons for the phase separation which could also be 'conducive' to the M–NM transition. While there is nothing in the Young–Alder discussion to point in that direction, there is a change in local order expected from discussions of concentrated electrolytes.

Kirkwood and Poirier (1954) have shown that there will be a transition from a Debye–Hückel structure to one showing oscillations in charge density similar to those in a fused salt (or ionic crystal) when the concentration of an electrolyte solution becomes sufficiently large that

$$\kappa_D b = 1 \cdot 03,$$

where κ_D is the Debye–Hückel screening wave number and b is the ionic radius. Stillinger and Lovett (1968a, b) have reached conclusions similar to Kirkwood and Poirier using an ion-pair theory; see also Stillinger, Kirkwood, and Wojtowicz (1960). Enderby (1974) has reported neutron diffraction experiments with appropriate structure in concentrated solutions of salts in water. While these theories do not encompass associations let alone clustering, so that the applicability is questionable, the change in local order might well allow for the changes seen in M–NH$_3$ solutions (Cohen and Thompson 1968).

As a consequence of the absence of structural data in the neighbourhood of the phase separation, and in view of the tentative nature of the models of phase separation in conducting fluids, it seems necessary to close this chapter with emphasis on the need for much more work. The areas in greatest need are: (1) X-ray or neutron diffraction; (2) critical exponents other than β and characterization of the crossover region; (3) critical point behaviour of transport coefficients; and (4) ionic solution theory for concentrated solutions. The conclusions which seem secure are: (1) the coincidence in the temperature–composition plane of the M–NM transition and the phase separation; and (2) the mean-field behaviour of the critical exponents for $\varepsilon \gtrsim 10^{-2}$.

REFERENCES

ADAMS, P. D. (1970). *Phys. Rev. Letts* **25**, 1012.
ANGELL, C. A. and SARE, E. J. (1968). *J. chem. Phys.* **49**, 4713.

ANTONIEWICZ, P. R. (1970). Private communication.

ASHCROFT, N. W. and LANGRETH, D. C. (1967). *Phys. Rev.* **155**, 682.

—— and LEKNER, J. (1966). *Phys. Rev.* **145**, 83.

BODDEKER, K. W. and VOGELGESANG, R. (1971). *Ber. (Dtsch) Bunsenges. phys. Chem.* **75**, 638.

BOWEN, D. E., THOMPSON, J. C., and MILLETT, W. E. (1968). *Phys. Rev.* **168**, 114.

BREDIG, M. A. (1964). In *Molten salt chemistry* (ed. M. Blander), Wiley and Sons, New York.

BRIDGES, R., INGLE, A. J., and BOWEN, D. E. (1970). *J. chem. Phys.* **52**, 5106.

BROUT, R. (1965). *Phase transitions.* Benjamin, New York.

CHIEUX, P. and SIENKO, M. J. (1970). *J. chem. Phys.* **53**, 566.

CHU, B., THIEL, D., TSCHARNUTER, W., and FENBY, D. V. (1972). *J. Phys. (Fr.)* **33**, 111.

CLARK, R. K. (1968). *J. chem. Phys.* **48**, 741.

—— and NEECE, G. A. (1968). *J. chem. Phys.* **48**, 2575.

COHEN, M. H. and JORTNER, J. (1974). *J. Phys. (Fr.)* **35**, C4–345.

—— and THOMPSON, J. C. (1968). *Adv. Phys.* **17**, 857.

CONIGLIO, A. (1972a). *Phys. Letts A* **38**, 105.

—— (1972b). *Physica* **58**, 489.

CORBETT, J. D. (1964). In *Fused salts* (ed. B. R. Sundheim), p. 341. McGraw-Hill, New York.

CUBICCIOTTI, D. D. (1949). *J. phys. Chem.* **53**, 1302.

D'ABRAMO, G., RICCI, F. P., and MENZINGER, F. (1972). *Phys. Rev. Lett.* **28**, 22.

DAMAY, P. and LEPOUTRE, G. (1971). *J. Chim. phys.* **68**, 970.

DEUTSCH, C. (1972). *J. non-cryst. Solids* **8–10**, 713.

DILLON, I. F., NELSON, P. A., and SWANSON, B. S. (1966). *J. chem. Phys.* **44**, 4229.

DOUMAUX, P. W. and PATTERSON, A. (1967a). *J. phys. Chem.* **71**, 3535.

—— —— (1967b). *J. phys. Chem.* **71**, 3540.

EGELSTAFF, P. A. and RING, J. W. (1968). In *Physics of simple liquids* (eds H. N. V. Temperley, J. S. Rowlinson, and G. S. Rushbrooke), p. 253. Wiley, New York.

—— and WIGNALL, G. D. (1970). *J. Phys, C (GB)* **3**, 1673.

ENDERBY, J. (1974). In *Amorphous and liquid semiconductors* (ed. J. Tauc), p. 361. Plenum Press, London.

EVEN, U. and JORTNER, J. (1973). In *Electrons in fluids* (eds J. Jortner and N. R. Kestner), p. 363. Springer-Verlag, Heidelberg.

FISHER, M. E. (1965). In *Lectures in theoretical physics* (ed. W. E. Brittin), p. 1. University of Colorado Press, Boulder, Colorado.

—— (1967). *Rep. Prog. Phys.* **30**, 615.

—— (1968). *Phys. Rev.* **176**, 257.

—— (1969). In *Contemporary Physics: Trieste Symposium 1968*, Vol. 1, p. 19. International Atomic Energy Agency, Vienna.

—— (1971). Theory of critical point singularities. In *Proceedings of the 1970 Enrico Fermi Summer School*, Course No. 51. Academic Press, New York and London.

—— and SCESNEY, P. E. (1970). *Phys. Rev. A* **2**, 825.

FREYLAND, W., PFEIFER, H. P., and HENSEL, F. (1974). In *Amorphous and liquid semiconductors* (eds J. Stuke and W. Brenig), p. 1327. Taylor and Francis, London.

GAMMELL, P. and ANGELL, C. A. (1972). National ACS Meeting (April 1972), Paper No. 154, Boston.

GREEN, M. S. (1971) (ed.). Critical phenomena. *Proceedings of the 1970 International Enrico Fermi Summer School*, Course No. 51. Academic Press, New York and London.

GRIFFITHS, R. B. and WHEELER, J. C. (1970). *Phys. Rev. A* **2**, 1047.

GSCHNEIDNER, K. A. (1964). *Solid St. Phys.* **16**, 275.

GUGGENHEIM, E. A. (1945). *J. chem. Phys.* **13**, 253.

HELLER, P. (1967). *Rep. Prog. Phys.* **30**, 731.

HENSEL, F. (1970). *Phys. Letts* **31A**, 88.

—— (1973). In *Electrons in fluids* (eds J. Jortner and N. R. Kestner), p. 335. Springer-Verlag, Heidelberg.

—— and FRANCK, E. U. (1968). *Rev. mod. Phys.* **40**, 697.

HOHENBERG, D. C. (1968). In *Conference on Fluctuations in Superconductors*, at Asilomar, California. Bell Telephone Labs, Murray Hill, New Jersey.

HSICH, S.-Y., GAMMON, R. W., MACEDO, P. B., and MONTROSE, C. J. (1972). *J. chem. Phys.* **56**, 1663.

ICHIKAWA, K. and THOMPSON, J. C. (1973). *J. chem. Phys.* **59**, 1680.

JASNOW, D., GOLDBURG, W. I., and SEMURA, J. S. (1974). *Phys. Rev. A* **9**, 355.

JOHNSON, J. W. and BREDIG, M. A. (1958). *J. phys. Chem.* **62**, 604.

KADANOFF, L. P., GOTZE, W., HAMBLEN, D., HECHT, R., LEWIS, E. A. S., PALCIAUSKAS, V. V., RAYL, M., SWIFT, J., ASPNES, D., and KANE, J. (1967). *Rev. mod. Phys.* **39**, 395.

KIKOIN, I. K. and SENCHENKOV, A. P. (1967). *Physics Metals Metallogr., N.Y.* **24**, 74.

KIRKWOOD, J. and POIRIER, T. C. (1954). *J. chem. Phys.* **58**, 591.

KRAUS, C. A. (1907). *J. Am. chem. Soc.* **29**, 1557.

KRUMHANSL, J. A. (1965). In *Physics of solids at high pressures* (eds C. T. Tomizuka and R. M. Emrick), p. 425. Academic Press, New York.

LEKNER, J. and BISHOP, A. R. (1973). *Phil. Mag.* **27**, 297.

LEPOUTRE, G. (1970). In discussion of Sienko, M. J., and Chieux, P., in *Metal-ammonia solutions* (eds J. J. Lagowski and M. J. Sienko), p. 339. Butterworths, London.

MOTT, N. F. (1961). *Phil. Mag.* **6**, 287.

—— (1974*a*). *Phil. Mag.* **29**, 613.

—— (1974*b*). *Metal–insulator Transitions*. Taylor and Francis, London.

NAIDITCH, S., PAEZ, O. A., and THOMPSON, J. C. (1967). *J. chem. Engng Data* **12**, 164.

NEECE, G. (1967). *J. chem. Phys.* **47**, 4112.

PITZER, K. S. (1958). *J. Am. chem. Soc.* **80**, 5046.

PREDEL, B. (1960). *Z. phys. Chem.* **24**, 206.

RENKERT, H., HENSEL, F., and FRANCK, E. U. (1971). *Ber. (Dtsch) Bunsenges. phys. Chem.* **75**, 507.

ROSS, R. G. (1971). *Phys. Letts* **34A**, 183.

SCHETTLER, P. D., DOUMAUX, P. W., and PATTERSON, A. (1967). *J. phys. Chem.* **71**, 3797.

—— and PATTERSON, A. (1964*a*). *J. phys. Chem.* **68**, 2865.

—— —— (1964*b*). *J. phys. Chem.* **68**, 2870.

SCHINDEWOLF, U. (1968). *Angew Chem.* **7**, 190.

——, LANG, G., and BODDEKER, K. W. (1969). *Z. phys. Chem.* **66**, 86.

SCHMUTZLER, R. and HENSEL, F. (1972). *J. Non-Cryst. Solids* **8–10**, 718.

SCHROEDER, R. L., THOMPSON, J. C., and OERTEL, P. L. (1969). *Phys. Rev.* **178**, 298.

SCHÜRMANN, H. K. and PARKS, R. D. (1971*a*). *Phys. Rev. Letts* **26**, 367.

—— —— (1971*b*). *Phys. Rev. Letts* **26**, 835.

SCHÜRMANN, H. K. and PARKS, R. D. (1971c). *Phys. Rev. Letts* **27,** 1790.

—— —— (1972). *Phys. Rev. B* **6,** 348.

SHARP, A. C., DAVIS, R. L., VANDERHOFF, J. A., LEMASTER, E. W., and THOMPSON, J. C. (1971). *Phys. Rev. A*, **4,** 414.

SIENKO, M. J. (1949). *J. Am. chem. Soc.* **71,** 2707.

—— (1964). In *Metal–ammonia solutions* (eds G. Lepoutre and M. J. Sienko), p. 23. Benjamin, New York.

STANLEY, H. E. (1971). *Introduction to phase transitions and critical phenomena.* Clarendon Press, Oxford.

STEIN, A. and ALLEN, G. F. (1972). *Phys. Rev. Letts* **29,** 1236.

—— —— (1973). *Phys. Rev. Letts* **59,** 6079.

STILLINGER, F. H., KIRKWOOD, J. G., and WOJTOWICZ, P. J. (1960). *J. chem. Phys.* **32,** 1837.

—— and LOVETT, R. (1968a). *J. chem. Phys.* **48,** 3858.

STROUD, D. (1974). *J. Phys. (Fr.)* **35,** C4–387.

SWIFT, J. B. (1973). *J. chem. Phys.* **58,** 5184.

TEOH, H., ANTONIEWICZ, P. R., and THOMPSON, J. C. (1971). *J. phys. Chem.* **75,** 399.

THOMPSON, J. C. (1967). In *Chemistry of non-aqueous solvents* (ed. J. J. Lagowski), Vol. 2, p. 265. Academic Press, New York.

—— (1968). *Rev. mod. Phys.* **40,** 704.

THOMPSON, D. R. and RICE, O. K. (1964). *J. Am. chem. Soc.* **86,** 3547.

WIDOM, B. (1962). *J. chem. Phys.* **37,** 2703.

—— (1973). *J. phys. Chem.* **77,** 2196.

WIGNALL, G. D. and EGELSTAFF, P. A. (1968). *J. phys. Chem.* **1,** 1088.

YAZAWA, G. and LEE, Y. K. (1970). *Trans. Jap. Inst. Metals* **11,** 411.

YOUNG, D. A. and ALDER, B. J. (1971). *Phys. Rev. A* **3,** 364.

ZOLLWEG, J. A. (1971). *J. chem. Phys.* **55,** 1430.

6

SOLUTIONS OF ALKALINE EARTH
AND RARE EARTH METALS

6.1. Introduction

DESPITE the fact that experiments on Ca–NH$_3$ solutions were reported by Mossian (1904) at the turn of the century the kind of detailed information which fills Chapters 2–5 has never been acquired for solutions of the alkaline earth and rare earth metals. This regrettable fact derives in part from the relative difficulty with which these metals are purified and handled (Rudolph 1971; Howell and Pytlewski 1969), but is also due to the general observation that there is not much difference in the behaviour of monovalent and divalent solutions. The solvated electron (Chapter 3) dominates the behaviour of the dilute solutions despite the obvious effect of a M^{2+} ion on all the association processes. The concentrated solutions are metallic. Even the effects of the magnetic moment carried by the Eu^{2+} ion have not been adequately explored. Indeed proof that Eu and Yb dissolve (Warf and Korst 1956) and occur in the divalent rather than trivalent state is rather recent in the history of metal–ammonia solutions. It is therefore possible to collect the known behaviour of both dilute and concentrated solutions of Ca, Sr, Ba, Eu, and Yb in NH$_3$, together with the metal–nonmetal transition, all into a single chapter. After a brief description of the phase diagrams, the remaining material will be organized in the same order as Chapters 2–4 and followed by a summary.

6.2. Phase diagrams

The phase diagrams of the solutions of divalent ions, insofar as they are known, qualitatively resemble those of the monovalent ions (Section 1.4) particularly when concentration is expressed as mole percent *electrons*, thereby accounting for the two valence electrons. This means, however, that phase separation begins with a lower metal content (0.04 m.p.m. at 213 K in Ca–NH$_3$)

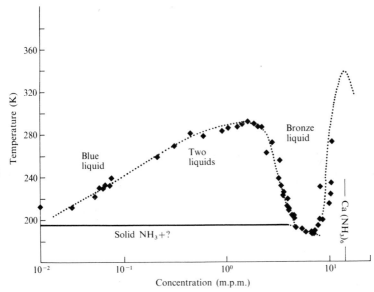

FIG. 6.1. Phase diagram of Ca–NH₃ solutions. The sources of data are quoted in the text. See the caption to Fig. 2.1 (p. 15) for the notation in this and subsequent figures.

than with the alkalis, and unwary experimenters have blundered into the two-phase region unknowingly. The solubility limits are also lower and have similarly been missed.

Fig. 6.1 shows the Ca–NH₃ diagram as determined by Okabe (1957), Hallada and Jolly (1963), Wong (1966); Schroeder, Thompson, and Oertel (1969); and Teoh, Antoniewicz, and Thompson (1971). The overall shape of the two-phase region is parabolic as discussed in Section 5.3; the compound $Ca(NH_3)_6$ is discussed in the next chapter along with the other hexammines. It is generally assumed that the consolute point is higher in Sr and Ba solutions though no experiments have been reported. Phase separation in Eu–NH₃ solutions has been observed but not studied.

At the normal boiling point of NH_3 one can only work with Ca in the ranges $0 \leqslant x \leqslant 0.0006$ and $0.05 \leqslant x \leqslant 0.09$. One must then surmount the difficulties presented by the rather narrow range of concentrations available in the normal liquid range of NH_3 as well as those problems listed in Section 6.1 if data are to be acquired.

6.3. Concentrated solutions

In solutions of divalent ions, concentrations in excess of the critical concentration, i.e. above about 3 m.p.m., will be taken as concentrated. There is too little information to warrant separating out an intermediate, or M–NM transition range as in Chapter 4.

6.3.1. Concentrated solutions: electronic properties

The metallic nature of concentrated solutions of divalent ions has been established by both conductivity (Wong 1966; Schroeder et al. 1969; Teoh et al. 1971; Vanderhoff and Thompson 1971) and Hall coefficient measurements (Kyser and Thompson 1965; Vanderhoff and Thompson 1971). The latter show two electrons per atom free and the former show conductivities as high as $4000 \, \Omega^{-1} \, cm^{-1}$, well above Mott's (1972) minimum conductivity. Fig. 6.2 shows results (Vanderhoff and Thompson 1971) for Ca–NH$_3$ solutions near 223 K together with similar data for Li– and K–NH$_3$ solutions, both plotted as functions of *valence* electron concentration. The close similarity requires displaced vertical scales so that the points do not overlap. Measurements of σ in Sr, Ba, and Yb solutions are similar to the Ca–NH$_3$ data (Schroeder et al. 1969). The one difference between the two kinds of solutes is the somewhat weaker dependence on concentration at higher electron concentrations in the divalent solutions (Schroeder et al. 1969). It appears that if one writes $\sigma \propto n^a$, where n is the number density of free electrons, then a is close to 3 in solutions of monovalent ions and close to 2 in solutions of divalent ions. The temperature coefficient $\gamma_T = \sigma^{-1} \, d\sigma/dT$ is about the same as in the alkalis (Fig. 2.5).

The optical constants (Somoano and Thompson 1970) are also consistent with two conduction electrons per atom though there are significant deviations from Drude (free-electron) theory. Since these are ellipsometric data they may not be accurate (see Section 2.2.5) and the differences may not be real. However, they do show noticeable differences from data taken by the same technique on solutions of monovalent metals.

A few susceptibility data have been reported by David, Glaunsinger, Zolotov, and Sienko (1973) as part of their work on solid hexammines.

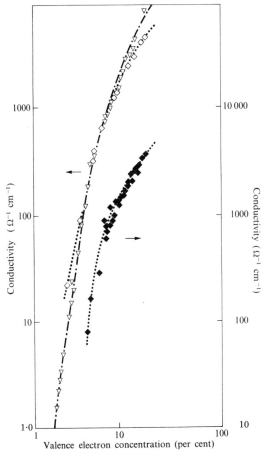

FIG. 6.2. Conductivity of several concentrated M–NH₃ solutions near 223 K (Vanderhoff and Thompson 1971). Note the difference in ordinate.

6.3.2. *Concentrated solutions: mechanical properties*

Extensive measurements of the density of Ca–NH₃ solutions in the concentrated range are contained in the unpublished thesis of Wa She Wong (1966). Indeed, these data were used in establishing the phase separation boundary of Fig. 6.1. The density data are shown in Fig. 6.3. The excess volume (see eqn 3.8) may be calculated and turns out to be more than twice that of the alkalis.

Were these solutions nonmetallic, one might attribute this to the two solvated electrons present per Ca^{2+} ion. However, the

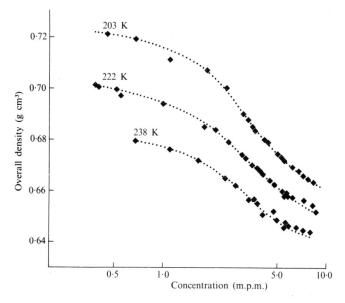

FIG. 6.3. Density of Ca–NH₃ solutions. Much of the data shown is in the two phase region (see Fig. 6.1) and therefore represents an average (Wong 1966).

electronic data clearly indicate that the electrons are free and cavities should not exist. Just as in the alkalis (Cohen and Thompson 1968; see also Section 2.2.9) the excess volume persists when delocalization occurs. The inhomogeneous model (Cohen and Jortner 1973a,b; Lelieur 1973) requires that some solvated electrons persist above the M–NM transition. The rapid decrease in ΔV, as metal is added to the Ca–NH₃ solution, suggests that the fraction of solvated electrons (if this is the correct explanation) is rapidly decreasing. Unfortunately there is not enough data to evaluate all the parameters of the inhomogeneous model.

Bridges, Ingle, and Bowen (1970) have reported sound speed measurements in calcium and barium solutions. A single point was reported by Maybury and Coulter (1951). In contrast to the monovalent solutions the sound speed shows a distinct minimum near 5 m.p.m. as may be seen in Fig. 2.19(a). It has been suggested that this minimum is related to the existence of the solid hexammine compounds inasmuch that a somewhat similar minimum is found in Li–NH$_3$ solutions but none is seen in Na– or K–NH$_3$ solutions (McAlister, Crozier, and Cochran 1973).

Vapour pressures have been measured by Marshall and Hunt (1956; and see early references therein) and by Okabe (1957). These data are mainly used to establish the existence of compounds (Chapter 7).

The unpublished viscosity measurements on Ca–NH$_3$ solutions of Earhart (1967) were made using a capillary viscometer and require both density and surface tension for interpretation. In the absence of the latter data, data for alkali metal solutions (Section 3.53) were used. The errors induced in the viscosity are presumably no more than a few per cent. The resulting viscosity is an increasing function of concentration, a trend opposite to that observed in Na–NH$_3$ solutions as may be seen in Fig. 6.4. Lower

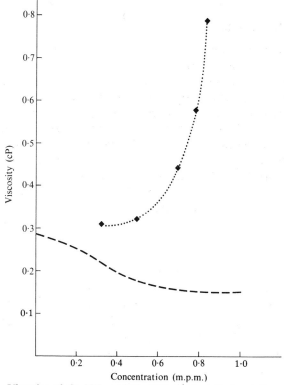

FIG. 6.4. Viscosity of Ca–NH$_3$ and Na–NH$_3$ (dashed line) solutions at 233 K (Earhart 1967). The abscissa is the *fraction* of saturated concentration.

temperatures result in a higher viscosity; the data are consistent with an Arrhenius relation.

6.3.3. *Concentrated solutions: summary*

The trends of density, sound speed, and viscosity are all consistent with an ordered fluid and it is plausible that some of the order of the solid hexammines persists into the liquid state. There is no doubt that both valence electrons are free and that mean free paths are sufficiently long for a metallic description to be justified at least near saturation. At lower concentrations, there are indications that an inhomogeneous model, comparable to that used for Na–NH$_3$ solutions (Lelieur 1973), is needed.

The inadequacy of the available data is clear. More data on solutions of Eu would be of considerable interest.

6.4. Dilute solutions

There is a distinct shortage of data here and many of the questions one might ask about the nature of the divalent solutions must be unanswered at present. Hallada and Jolly (1963) conclude from the optical absorption line in dilute Ca–NH$_3$ solutions that both valence electrons are solvated. Burow and Lagowski (1965) have the same conclusion for Ba–NH$_3$ solutions. The shape and location of the absorption is identical to that of alkali solutions (Fig. 3.18). The molar extinction coefficient is about twice that of the monovalent solutions. Freed and Sugarman (1943) similarly concluded that both valence electrons are solvated from a few measurements of susceptibility in Ca– and Ba–NH$_3$ solutions, though they found some evidence of spin pairing.

6.4.1. *Dilute solutions: electronic properties*

Conductance measurements have been made with dissolved calcium by Wong (1966) and with barium by Rudolph (1971). Neither has appeared in the journal literature. Fig. 6.5 shows the data near 210 K (Rudolph took data only at 211 K) compared to that for Na–NH$_3$ solutions. Despite the fact that electron densities are much higher in the alkaline earth solutions, if the atoms are doubly ionized, the conductivities (conductances) are much lower. The association processes are doubtless much

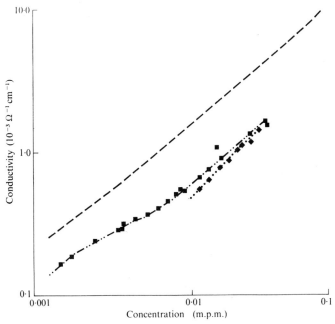

FIG. 6.5. Conductivity of dilute M–NH₃ solutions (Wong 1966; Rudolph 1971). The dashed line shows data for Na–NH₃ solutions.

further along than at the same alkali concentration because of the extra ionic charge.

The temperature cofficient $\sigma^{-1}\,d\sigma/dT$ varies from 2.7 to 4.25 per cent per degree (Wong 1966) with sufficient error to preclude assignment of a concentration dependence. This result exceeds that due to viscosity alone but all data point toward association. One may thus assign most of the observed temperature variation to the change of some equilibrium with temperature.

The Eu^{2+} ion has its own e.s.r. absorption and an optical absorption as well. Therefore one may look for the influence of the unpaired spin of the Eu^{2+} upon the unpaired spin of the solvated electron (Catterall and Symons 1965a,b; Thompson, Schaefer, and Waugh 1966; Thompson, Hazen, and Waugh 1966; Catterall 1970). The optical spectra remain unchanged in form as the concentration is increased while the spin resonance spectra are broadened—presumably by the formation of species involving both solvated cations and anions such as ion-pairs. The spin pairing of solvated electrons occurs without any noticeable effect on the Eu^{2+} spectra.

Cutler and Powles (1962) found T_1 and T_2 equal in e.s.r. experiments at 23 °C in Ca–NH₃ solutions. The value of 3.5 μsec is quite close to their results for Li, Na, and K in NH₃. Since the temperature was so high phase separation did not occur. They found T_1 (and T_2) to decrease as the metal concentration increased above 3×10^{-2} m.p.m.; a similar decrease occurs only near 10^{-1} m.p.m. (see Section 3.7, Section 4.2) in solutions of alkali metals, and is a part of the metal–nonmetal transition there.

6.4.2. *Dilute solutions: other properties*

There seem to have been no other reliable data reported at concentrations below that of the phase separation. Densities etc. are sufficiently close to the values for pure NH₃ that the difficult measurements have not been attempted, or else lurk in unpublished theses.

6.5. Metal–nonmetal transition

There have been no direct studies of the M–NM transitions in any solutions containing divalent ions. Cutler and Powles (1962), as noted in Section 6.4.1, may have seen signs of electron delocalization in the drop of the e.s.r. T_1; Vanderhoff and Thompson (1971) saw an increasing Hall coefficient as they decreased concentrations (Section 4.2); and Teoh *et al.* (1971) assembled the conductivity data shown in Fig. 6.6 from their phase separation observations. The data were taken quite close to the consolute point (Chapter 5) and reveal a much sharper transition than is seen in alkali metal solutions.

Clearly no analysis of the sort carried out by Lelieur (1973) or Cohen and Jortner (1974) can be made here. Nevertheless, it seems highly probable that such an analysis might reproduce the behaviour of these solutions as well as those with monovalent ions. The increased ionic charge of the M^{2+} appears to shift the transition to lower concentrations, and lowers the mobility in the dilute fraction but otherwise the transition is likely to be dominated by the solvated and free electrons.

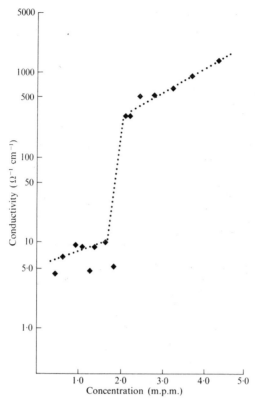

FIG. 6.6. Conductivity changes near the consolute point in Ca–NH₃ solutions (Teoh *et al.* 1971).

6.6. Summary

The constructs which have served for solutions of alkalis appear to suffice for the alkaline and rare earths as well. While the details of ion-pairing and spin-pairing are not revealed by the meager data reported here, the isolation of the solvated electron(s) has been beautifully confirmed by the e.s.r. measurements. The notion of weak cation involvement in spin-pairing processes and the likelihood of two-electron cavities is enhanced by these observations.

Finally it should be noted that an even wider variety of positive ions can be created in NH₃ through the use of electrochemical techniques (Jolly 1959; Quinn and Lagowski 1968). The species observed range from aluminium to tetraalkylammonium radicals. No metals other than those mentioned can be *dissolved.*

REFERENCES

BRIDGES, R., INGLE, A. J., and BOWEN, D. E. (1970). *J. chem Phys.* **52,** 5106.
BUROW, D. F. and LAGOWSKI, J. J. (1965). *Adv. Chem. Ser.* **50,** 125.
CATTERALL, R. (1970). In *Metal–ammonia solutions* (eds J. J. Lagowski and M. J. Sienko), p. 105. Butterworths, London.
—— and SYMONS, M. C. R. (1965a). *J. chem. Phys.* **42,** 1466.
—— —— (1965b). *J. chem. Soc.* 3763.
COHEN, M. H. and JORTNER, J. (1973a). *Phys. Rev. Letts* **30,** 696.
—— —— (1973b). *Phys. Rev. Letts* **30,** 699.
—— —— (1974). *J. Phys. (Fr.)* **35,** C4–345.
—— and THOMPSON, J. C. (1968). *Adv. Phys.* **17,** 857.
CUTLER, D. and POWLES, J. G. (1962). *Proc. phys. Soc. Lond.* **80,** 130.
DAVID, T., GLAUNSINGER, W., ZOLOTOV, S., and SIENKO, M. J. (1973). In *Electrons in fluids* (eds J. Jortner and N. R. Kestner), p. 323. Springer-Verlag, Berlin, Heidelberg, and New York.
EARHART, J. P. (1967). M.S. Thesis, University of California at Berkeley (Unpublished).
FREED, S. and SUGARMAN, N. (1943). *J. chem. Phys.* **11,** 354.
HALLADA, C. J. and JOLLY, W. L. (1963). *J. inorg. chem.* **2,** 1076.
HOWELL, K. and PYTLEWSKI, L. L. (1969). *J. less-common Metals* **19,** 399.
JOLLY, W. L. (1959). *Prog. inorg. Chem.* **1,** 235.
KYSER, D. S. and THOMPSON, J. C. (1965). *J. chem. Phys.* **42,** 3910.
LELIEUR, J.-P. (1973). *J. chem. Phys.* **59,** 3510.
MARSHALL, P. and HUNT, H. (1956). *J. phys. Chem.* **60,** 121.
MAYBURY, R. H. and COULTER, L. V. (1951). *J. chem. Phys.* **19,** 1326.
MCALISTER, S. P., CROZIER, E. D., and COCHRAN, J. F. (1973). *J. Phys. C (GB)* **6,** 2269.
MOISSAN, H. (1904). *Bull. Soc. chim. Paris* **31,** 549.
MOTT, N. F. (1972). *Phil. Mag.* **26,** 1015.
OKABE, T. (1957). *J. Soc. chem. Ind. Japan* **60,** 1438.
QUINN, R. K. and LAGOWSKI, J. J. (1968). *J. phys. Chem.* **72,** 1374.
RUDOLPH, M. J. (1971). Ph.D. Dissertation, Drexel University. (Unpublished).
SCHROEDER, R. L., THOMPSON, J. C., and OERTEL, P. L. (1969). *Phys. Rev.* **178,** 298.
SOMOANO, R. B. and THOMPSON, J. C. (1970). *Phys. Rev. A* **1,** 376.
TEOH, H., ANTONIEWICZ, P. R., and THOMPSON, J. C. (1971). *J. phys. Chem.* **75,** 399.
THOMPSON, D. S., HAZEN, E. E., and WAUGH, J. S. (1966). *J. chem. Phys.* **44,** 2954.
——, SCHAEFER, D. W., and WAUGH, J. S. (1966). *Inorg. Chem.* **5,** 325.
VANDERHOFF, J. A. and THOMPSON, J. C. (1971). *J. chem. Phys.* **55,** 105.
WARF, J. C. and KORST, W. L. (1956). *J. phys. Chem.* **60,** 1590.
WONG, W. S. (1966). Ph.D. Dissertation, University of California at Berkeley. (Unpublished).

SOLID METAL–AMMONIA COMPOUNDS

7.1. Introduction

THE solid compounds formed at low temperatures between ammonia and most of the metals soluble in ammonia can be understood only in the framework of solid state physics (*cf.*: Mott and Jones 1936; Ziman 1964) not otherwise a part of the rest of this book. However, once understood these compounds may well shed light on the nature of the M–NH₃ solutions. This chapter therefore is included.

M–NH₃ compounds characteristically retain the golden colour of the concentrated liquid solutions. The lithium compound has the lowest melting point (near 90 K) of any metal as well as a wide variety of striking transport properties. The europium compound also continues the rich variety of physical properties associated with the M–NH₃ system. Elucidation requires detailed application of sophisticated solid state theory and experiment.

Kraus, whose data have been quoted often before in this monograph, concluded in 1908 from vapour pressure measurements that stable compounds on the form Ca(NH₃)₆ should exist. Jaffe, in 1935, studied the solid formed when a saturated lithium–ammonia solution was frozen and found a Hall coefficient consistent with one free electron per metal atom. He also measured the magnetoresistance at the temperature of both liquid nitrogen and liquid hydrogen. The greatest impetus to work on frozen metal–ammoniates or metal–amines came, however, from Ogg's announcement in 1946 (Ogg 1946*a*) that a frozen sodium–ammonia solution was a superconductor at 77 K. Though no other experimenter (e.g. Boorse, Cook, Pontius, and Zemansky 1946) was able to reproduce his results, considerable information about the frozen solution was obtained in the course of the effort.† These and more recent results will be discussed in this chapter.

Not all of the metals known to dissolve in liquid ammonia form stable solids and of course the composition of the solids that do form is limited. Sodium and potassium are known to precipitate as metals when the solution is frozen. Lithium, calcium, strontium, barium, europium, and ytterbium form solid compounds and the lattice structure and unit cell dimensions are known. Though the solid formed on freezing Cs–NH₃ retains the characteristic gold or bronze colour of the liquid (in contrast to the white, colloidal suspension obtained with Na or K) neither the structure nor the formula is known. Indeed there are other data which indicate *no* compound of Cs with NH₃. The present author and his students have studied transport coefficients and heat capacity. Mammano, who is responsible for some of the X-ray work, has also done heat capacity work. Varlashkin has studied the electron momentum distribution with positrons, as in the liquid, and Levy has measured the e.s.r. spin–lattice relaxation time. Sienko and his co-workers have contributed a variety of data. The alkali, alkaline earth, and rare earth compounds will be discussed in that order. Within each grouping structural data will precede other properties and an attempt at a model will be made.

A recent, exciting paper (Dye, Ceraso, Lok, Barnett, and Tehan 1974) reports an entirely new class of golden compounds. These combine Na with ethylamine rather than NH₃ (see Chapter 8) and a cation-complexing ether (crypt). The results to date are meager but the composition is $Na_2C_{18}H_{36}N_2O_6$ and the structure is hexagonal. It appears that the material is semiconducting rather than metallic as in the M–NH₃ compounds. One unexpected point is that the two sodium species in each structural unit are different: one is trapped in the crypt and the other is outside;

---

† It is impossible at this late date to establish the origin of Ogg's mistake. The last of his letters to the Editor in Physical Review (Ogg 1946*b*) on the subject concluded with the statement '... indicates a resistance of 10^{-13} Ω or smaller. In view of the fact that the liquid samples possessed resistances of the order of a thousand ohms, the solids in question may fairly be said to be superconducting.' Those who made contemporary attempts to reproduce his results were generally unable to detect signals other than those due to other apparatus in the laboratory. More interesting even than the experiment was the theory which Ogg proposed to explain the results. The fundamental assumption involved Bose–Einstein condensation of electrons with paired spins. Theory and experiment were both quickly discounted. However, a recent Soviet report re-opens the issue (Dmitrenko and Shchetkin 1973).

the trapped species is apparently Na^+ while the counter-ion is Na^-. It is likely that other compounds of this type can be prepared.

7.2. Li(NH₃)₄

7.2.1. *Phase diagram*

The compounds of metal with NH_3 are formed by dissolving the metal in liquid ammonia then freezing the resulting solution. In the case of Li the compound composition (20 m.p.m.) is quite close to that of the eutectic. There is, therefore, considerable controversy over the precise form of the phase diagram (Morgan, Schroeder, and Thompson 1965; Mammano and Coulter 1967; Mammano 1970; David, Glaunsinger, Zolotov, and Sienko (1973). It is generally agreed that there is a transition near 88·8 K (Mammano and Coulter 1969; David *et al.* 1973) another in the solid at 82·4 K and some peculiar thermal effects in the 65–70 K range. The eutectic is variously reported at 19·4 and 20 m.p.m. (Mammano and Coulter 1967; David *et al.* 1973). The best current phase diagrams (See also Fig. 1.1) are shown schematically in Fig. 7.1 and were based on DTA studies (David *et al.* 1973) of a wide variety of Li–NH₃ samples.

The DTA traces showed three peaks. Near 88·5 K was a peak

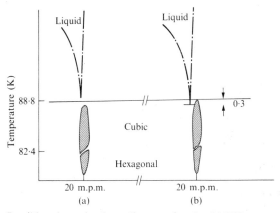

FIG. 7.1. Possible schematic phase diagrams for the Li–NH₃ system near the 89 K and 82 K transitions (David *et al.* 1973). The eutectic is marked at 19·4 m.p.m.

which on occasion showed as much as $0 \cdot 3$ K splitting. As noted in the figure, one must therefore assume a eutectic at $88 \cdot 8$ K and $19 \cdot 4$ m.p.m. with a peritectic at $88 \cdot 5$ and 20 m.p.m. *or* else invert the temperatures assigned to eutectic and peritectic. The quadruple point once postulated for the 88 K event (Mammano 1970) has been ruled out by the observation of a substantial pressure shift in its temperature: $0 \cdot 01$ K atm^{-1}. The low temperature phase is cubic (See Section 7.2.2).

The DTA peak at $82 \cdot 4$ K or $82 \cdot 2$ K (Mammano and Coulter 1967) is assigned to the cubic-to-hexagonal solid–solid phase transition observed on cooling. The slight splitting observed in this peak has *not* received an explanation as yet. The transition shows a $0 \cdot 038$ K atm^{-1} pressure shift.

A peak was also observed near 69 K when excess Li was present in the sample and was assigned by David *et al.* to the well-known Martensitic transition in pure Li. For concentrations below 20 m.p.m. the 69 K peak was replaced by a peak near 196 K due to the melting of excess NH_3. A number of workers (Morgan *et al.* 1965; Cate and Thompson 1971; LeMaster and Thompson 1972; also Rosenthal and Maxfield 1973) have reported anomalies and hysteresis in various parameters in this same range. Other data show minor anomalies in this range also (Glaunsinger, Zolotov, and Sienko 1972) while the precise heat capacity measurements of Mammano and Coulter (1969) show nothing unusual. While the DTA results suggest that excess Li may be involved, resistivity anomalies were observed even in the presence of excess NH_3 (Rosenthal and Maxfield 1973). The peculiar effects remain to be explained and the existence of a third phase below 65 K cannot be ruled out. Parker and Kaplan (1973) suggest a cessation of rotation of NH_3 molecules about the dipole axis as a possible cause for an anomaly in $Eu(NH_3)_6$; perhaps a similar mechanism is responsible for the effects reported here *if* they are real.

$Li(NH_3)_4$ thus has two known solid phases: cubic between 82 and 88 K, and hexagonal below 82 K. Low temperature X-ray or neutron studies will be required to resolve the origin of the 65–70 K transition. The presence of the solid–solid phase transition precludes the production of single crystals for the low temperature studies required to resolve some of the questions on electronic structure raised below.

7.2.2. *Structure*

Structural studies have been reported by Mammano and Sienko (1968) and by Kleinman, Hyder, Thompson, and Thompson (1970). Neither is definitive in that temperature control was inadequate and the data taken below 82 K showed the presence of both phases. It nevertheless seems clear that the phase existing between 82 and 88 K is cubic with $a = 9·55$ Å. A density of $0·57$ gm cm^{-3} may be calculated for four molecules per unit cell. The hexagonal phase has $a = 7·0$ Å and $c = 11·1$ Å; the c/a ratio is thus $1·58$ in contrast to the ideal value of $1·633$. The density is presumed to be $0·53$ gm m^{-3}, but other considerations (section 7.2.4) suggest this value is low. A possible structure for hexagonal $Li(NH_3)_4$ is shown in Fig. 7.2. The different shadings indicate NH_3 molecules at different levels along the c-axis while the Li^+ ions are located at the centres of tetrahedra underneath the points labeled 1. The assumed radius for the MH_3 molecule is $1·657$ Å and that for the Li^+ ion is $0·60$ Å. This structure is consistent with the available X-ray data but must not be taken as

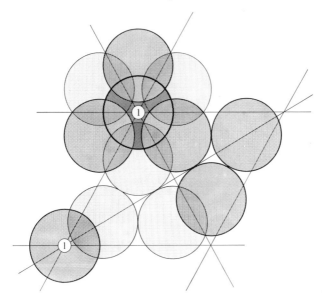

FIG. 7.2. Possible structure for hexagonal phase $Li(NH_3)_4$. Li^+ ions are located at the points marked 1, and the NH_3 molecules are arranged tetrahedrally about the ions. The lightest stippling denotes the lowest phase of NH_3 molecules, the darkest stippling the next, and the medium is on top.

final. Sienko (1964) has described the $Li(NH_3)_4$ molecule as consisting of NH_3 molecules tetrahedrally arranged about the Li^+ ion. The 2s electron of the Li atom is assumed promoted to a 3s orbital (and the conduction band). The 2s2p orbitals are then hybridized into sp^3 states which are occupied by the lone pair electrons from the NH_3 molecules. The 3s orbital is thought to be primarily outside of the NH_3 molecular sheath. The 3s state is then the basis for the conduction band.

7.2.3. Energetics

The energetics of formation of each of the metal–ammonia compounds can be analyzed in terms of a Born–Haber-like analysis using reactions of the sort (Mammano 1970):

$$Li(s) + 4NH_3(1) \rightarrow Li(NH_3)_4(s), \qquad (7.1)$$

where the suffixes (s) (1), and (g) refer to solid, liquid and gas, respectively. The various steps in the reaction are visualized as follows:

(a) metal converted from solid to gas (sublimated);
(b) metal ionized;
(c) ammonia converted from liquid to gas;
(d) gas phase metal ion is ammoniated;
(e) ammoniated metal recaptures electron;
(f) ammoniated metal converted from gas to molecular solid;
(g) electrons in solid delocalized.

The sum of the processes above is equivalent to the reaction of eqn (7.1). Most of the steps (a)–(g) can be evaluated only approximately and theoretically (Mammano and Sienko 1968; Mammano 1970; see also Section 7.3.2). See Table 7.1. Step (a) is the sublimation of the metal and uses the known heat of sublimation; the energy in step (b) is the first ionization potential of the gaseous atom; the latent heat of boiling for NH_3 enters in step (c). For step (d), the bonding energy of the ion complex in the gas phase is taken to be the difference in solvation energies between the bare and dressed ion. The former is determined from experimental data: namely the free energy of solution $\Delta G = -9.2$ kcal obtained by Marshall (1964) from vapour pressures and the heat of solution $\Delta H = -39$ kcal for the solvated electron heat of solution quoted by Jolly (1959). This overestimates the gas-phase ammoniation energy by the solvation energy of the ion complex or dressed ion which is estimated (Mammano and

TABLE 7.1

Energy contributions to $M(NH_3)_n$ *formation* (*in* eV)

Process	$M = Li$[a] $n = 4$	Eu[b] 6	Yb[b] 6	Sr[b] 6
a $M(s) \rightarrow M(g)$	+1·61	1·87	1·86	1·69
b $M(g) \rightarrow M^{n+}(g) + ne^-$	+5·36	17·03	18·35	16·67
c $n(NH_3)(1) \rightarrow n(NH_3)(g)$	+1·22	1·82	1·82	1·82
d $M^{n+}(g) + n(NH_3)(g) \rightarrow M(NH_3)_n^{n+}(g)$	−3·12	−9·5	−10·1	−8·9
e $M(NH_3)_n^{n+}(g) + ne^- \rightarrow M(NH_3)_n(g)$	−1·91	−8·02	−8·15	−7·15
f $M(NH_3)_n(g) \rightarrow M(NH_3)_n$ (*molec. solid*)	−0·27	−0·94	−1·02	−0·70
g $M(NH_3)_n$ (*molec. solid*) $\rightarrow M(NH_3)_n$ (*metal*)	−4·04	−4·04	−4·13	−4·04
Net $M(s) + n(NH_3)(l) \rightarrow M(NH_3)_n(s)$	−1·15	−1·78	−1·37	−0·61

[a] Mammano and Sienko (1968). [b] Oesterreicher, Mammano, and Sienko (1969).

Sienko 1968) to be −51 kcal. The net result of −72 kcal mol^{-1} is then shown in Table 7.1. The addition of an electron, so as to make the dressed ion neutral, is taken as the difference of the $2s \rightarrow 3s$ transition and the $2s \rightarrow$ continuum energy for Li and yields the energy of step (*e*). The condensation of the molecular gas into a molecular solid yields energy (*f*) estimated as near zero following Oesterricher, Mammano, and Sienko (1969). The delocalization energy step (*g*), is computed from a Hartree model with Pines's estimate of the correlation energy and is quite large (Mammano and Sienko 1968). The result is that the solid $Li(NH_3)_4$ exists *stably* primarily as a result of the energy gain produced by the delocalization of the Li valence electron. The gas phase 'molecule' on the other hand is unstable by about 0·34 eV in this approximation.

7.2.4. *Heat capacity*

The arguments in the preceding paragraph reinforce the structural data which indicate that there is actually a stable solid of the stoichiometry $Li(NH_3)_4$. Further support comes from the heat capacity measurements of Mammano and Coulter (1967, 1969). They made runs on the solid frozen from a number of different Li–NH$_3$ solutions. Reaction of the solutions was retarded by enclosing the sample in a polyethylene envelope. Other calorimetric techniques were routine. Contrary to other experimenters, the 82·2 K transition was found to be slow. Data for a 22·25 m.p.m. sample are shown in Fig. 7.3. The first point is that the heat capacity of the solid is significantly (20–50 per cent) higher than that of the individual components and also the

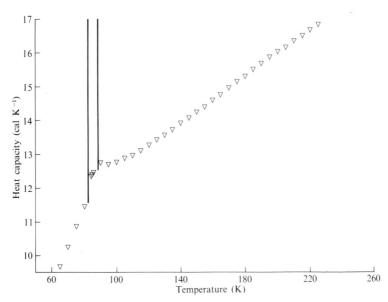

FIG. 7.3. Heat capacity of Li(NH₃)₄ in solid and liquid phases. The two vertical lines mark phase transitions, see Fig. 7.1 (Mammano and Coulter 1969). See the caption to Fig. 2.1 (p. 15) for the notation in this and subsequent figures.

Dulong–Petit value.† The latent heats are $520\cdot3\pm0\cdot7$ cal mol⁻¹ and $553\cdot2\pm1\cdot1$ cal mol⁻¹ at the $82\cdot2$ and $88\cdot8$ K transition, respectively. The transitions appear to be first order. Mammano found no thermal anomaly in the 60–70 K range, even with excess Li.

The enthalpy change associated with each transition, at different concentrations, varies according to the amount of excess Li or excess NH₃. In order to obtain internal consistency in their data, Mammano and Coulter (1967) were forced to assign a composition of 19.4 m.p.m. to the eutectic. The DTA results are not inconsistent with this result.

The DTA measurements of David *et al.* (1973) also contribute calorimetric information. They have measured the shift in each transition with pressure, up to 30 bar, and their results are shown in Fig. 7.4. Once $\partial T/\partial P$ is known the Clausius–Clapeyron equation permits the latent heats to be used to compute the volume changes at the two transitions. Since $\partial T/\partial P$ is positive at both transitions the

† The data above $88\cdot8$ K are, of course, for the liquid plus a small excess of metallic Li.

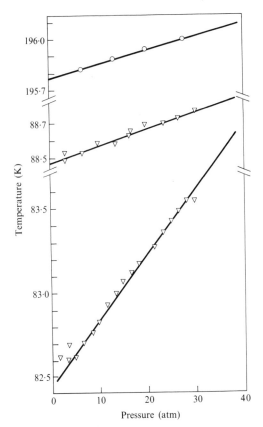

FIG. 7.4. DTA determination of the pressure shift of the melting point of NH_3 (top curve), melting point of $Li(NH_3)_4$ (middle), and the hexagonal–cubic phase transition (lower curve) (David *et al.* 1973).

density of the low temperature phase must exceed that of the high temperature phase.

Starting with the measured density (Mammano and Sienko 1968) of the hexagonal phase at $0·57 \, gm \, cm^{-3}$, the density of the cubic phase becomes $0·53 \, gm \, cm^{-3}$ and that of the liquid $0·52 \, gm \, cm^{-3}$ (Lo 1966). These results are in disagreement with the X-ray results, but are much more reliable.

Morgan and Thompson (1967) have reported low temperature heat capacity data. Their samples were enclosed in a Cu calorimeter and showed little decomposition when kept cold.

While they were not able to go low enough to see a temperature-independent Debye temperature Θ_D they quote a value of 55 K. Electronic contributions to the heat capacity could not be resolved.

7.2.5. *Transport coefficients*

The results of electrical measurements on Li(NH₃)₄ are the source of most of the current interest in this material. The electrical resistivity ρ, particularly the magnetoresistance, is the most striking property of Li(NH₃)₄. Fig. 7.5 shows a Koehler diagram based on data obtained at both helium and hydrogen temperatures (Jaffe 1935; McDonald and Thompson 1966; Rosenthal and Maxfield 1973). Note that, at 1·66 K, $\Delta\rho/\rho_0 =$ 5000 in a field of 100 kG. In most polycrystalline, uncompensated metals the values of $\Delta\rho/\rho_0$ are one or two orders-of-magnitude lower. A systematic presentation of resistivity Hall coefficient, thermoelectric power and optical data will be given before the magnetoresistance data are analysed in terms of a Fermi surface consistent with the structure. The cubic phase will be discussed before the hexagonal.

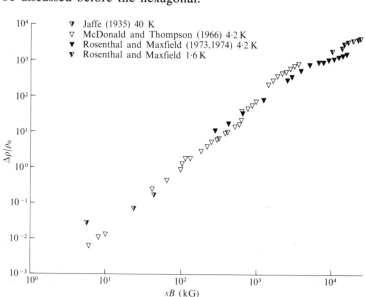

FIG. 7.5. Kohler diagram of magnetoresistance in Li(NH₃)₄. s^{-1} is the ratio of the resistivity at the temperature in question to that at 77 K.

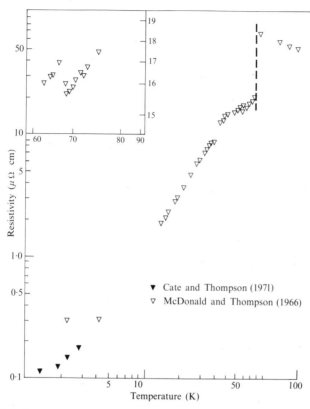

FIG. 7.6. Temperature dependence of the resistivity of $Li(NH_3)_4$. The melting point is marked by a vertical dashed line. The inset shows the disputed phase change near 67 K.

The resistivity data (Morgan, Schroeder, and Thompson 1965) in Fig. 7.6 show the resistive changes on freezing and the cubic–hexagonal change (ratios of 6·5 and 1·3, respectively, are observed) together with the anomalies observed near 70 K. No superconductivity has been observed. In the 1–20 K region there is a T^2 dependance which dominates the usual T^5 term (Cate and Thompson 1971). Unpublished data (Maxfield, private communication) show that the temperature dependent resistivity continues well below 1 K before reaching an impurity limit. While the overall trends above 20 K are qualitatively close to the Bloch–Gruneisen (phonon scattering) form (Morgan and Thompson

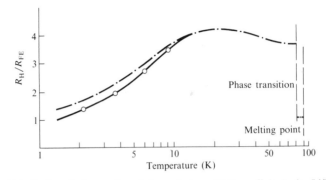

FIG. 7.7. Ratio of observed and free electron Hall coefficients in $Li(NH_3)_4$ (LeMaster and Thompson 1972). The two lines below 10 K indicate limits of error.

1967) they are significantly different because of Umklapp effects, as pointed out by Rosenthal and Maxfield (1973).

LeMaster and Thompson (1972) have reported Hall effect and thermopower data, which supplement the old Hall data of Jaffe (1935). Fig. 7.7 shows some of their results. The data were taken with a double a.c. technique and included contributions from the magnetoresistance. Corrected data are shown along with the original results, but the former may be in error by as much as 100 per cent at the lowest temperatures.

Fig. 7.8 shows the absolute thermoelectric power measured against Pb, there is a sign change near 25 K as well as the usual anomalies near 70 K and the discontinuities at 82 and 89 K. The authors suggest that the effects due to polycrystallinity and grain boundaries are less for the Hall and Seebeck voltages than in resistivity measurements (Herring 1960). There is clearly less variation from sample-to-sample and with different thermal histories.

Optical data have been taken on the solid compound by Vanderhoff and by McKnight but are unpublished. The golden colour persists in both phases. The apparatus used was the same as that used by Mueller and Thompson (1969) (Mueller 1969). The data contain errors intrinsic to the reflectance technique. The optical properties of the cubic phase are quite similar to those of the liquid, despite the 7-fold change in d.c. conductivity between the two phases. In the hexagonal phase, there is an extra absorption in the 1–2 eV range which may be due to interband effects

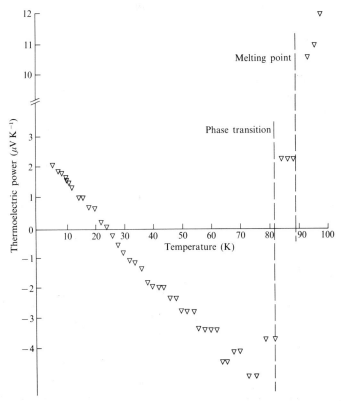

FIG. 7.8. Thermoelectric power of Li(NH₃)₄ (LeMaster and Thompson 1972).

(see below). More data, particularly at low energies, is certainly required.

As noted in the beginning of this chapter, it is the magnetoresistance which has drawn the most attention to $Li(NH_3)_4$. The observed effect is large, hasn't completely saturated at 100 kG, and doesn't obey Kohler's rule at the lowest temperatures (Rosenthal and Maxfield 1973, 1974). The latter effect is best seen in Fig. 7.5 where at high fields the resistivity obeys a $B^{0.6}$ power law while the exponent is 1·5 at low fields; however, the change in power law occurs at different values of B/ρ_0 on the Kohler diagram. For fields below 1 kG, LeMaster and Thompson report a power law of the form $B^{1.0}$, Rosenthal and Maxfield did not cover this range. Interpretation of these as well as other properties is made difficult if not impossible by the fact that all low temperature samples are polycrystalline.

7.2.6. *Magnetic data*

The earliest magnetic data were esr measurements reported by Levy (1956). His interpretation was rendered ambiguous by confusion over the phase of the system and poor temperature measurement. It seems clear nevertheless that the e.s.r. line width is an increasing function of temperature in what is now known to be the hexagonal phase, in contrast to the temperature independent line width in pure lithium. More recent measurements have been reported briefly by David *et al.* (1973).

Glaunsinger *et al.* (1972) have made careful susceptibility measurements using the Faraday method over the 194 K range. Their liquid state results are consistent with the paramagnetism and slight increase with T discussed in Chapter 2. As may be seen in Fig. 7.9 there is a slight decrease on freezing followed by a 25 per cent decrease at the cubic-to-hexagonal transition. Curie–Weiss behaviour is observed in the hexagonal phase down to 15 K with a flattening below that point. In some runs slight anomalies were observed in the 60–70 K range.

7.2.7. *Other data*

Varlashkin and Arias-Limonta (1971) has measured the momentum distribution of positrons annihilating in $Li(NH_3)_4$ *near* 77 K.

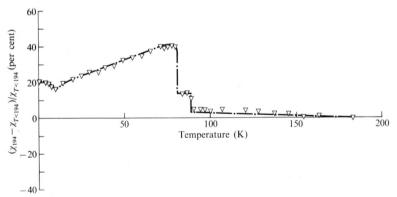

FIG. 7.9. Difference between observed susceptibility of $Li(NH_3)_4$ and that at 195 K, with the effect of density changes also subtracted out (Glaunsinger *et al.* 1972).

When all data are normalized to the same peak height (at zero angle) the $Li(NH_3)_4$ results are close to those of the liquid. In view of the general insensitivity of the positron to concentration in the liquid (see Chapters 2, 3, and 4) no real information can be derived from the observation.

No other observations have been reported.

7.2.8. A model

Knowledge of the geometries of Brillouin zone and Fermi surface is the starting point for understanding the electronic properties of any metal (Harrison 1970). In the absence of single crystals, detailed description of $Li(NH_3)_4$ is impossible. Nevertheless an attempt will be made to relate the observed electronic properties to general notions about the Fermi surfaces in f.c.c. and h.c.p. crystals.

The face-centred cubic phase ($82 < T < 89$ K) contains one Li atom per primitive cell and the Li contributes one electron per atom to the conduction band. The shortest distance from the center of the Brillouin zone to its face is in the $\langle 111 \rangle$ direction is $5 \cdot 7 \times 10^7$ cm^{-1} in this system. The free electron Fermi sphere has a radius $k_F = 5 \cdot 1 \times 10^7$ cm^{-1} and must therefore be contained entirely within the first Brillouin zone. The magnetic data of Glaunsinger et al. (1972) may be analyzed in terms of the standard Pauli theory to yield an effective mass ratio of $m^*/m = 1 \cdot 83$. This result is consistent with no zone contact but does suggest that the Fermi surface is non-isotropic. Bulging in the $\langle 111 \rangle$ direction is the most likely.

The absence of appreciable anisotropies in the Fermi surface is consistent with the observed transport properties of the cubic phase of $Li(NH_3)_4$.

The seven-fold change in resistivity on freezing may be explained, following Mott, as simply resulting from the increase in ordering as the solid forms (Rosenthal and Maxfield 1973). This is supported by the observation that the change in susceptibility across the freezing point can be accounted for in terms of the density change (Glaunsinger et al. 1972). The smallness of the change (-6 per cent) indicates that the density of states is only slightly changed on freezing. The Hall coefficient is also in good agreement with the quasi-free model.

The thermopower, however, has a positive sign as opposed to

the negative sign predicted for both the diffusion and phonon-drag contributions to the thermopower. In view of the fact that many of the alkali and noble metals have negative Hall coefficients and positive thermopowers, the behaviour seen here is not surprising (LeMaster and Thompson 1972).

In short, the cubic phase of tetraamminelithium is a well-behaved and fairly simple metal. Nothing could be more different than the hexagonal phase. In the first place, there are only two other h.c.p. monovalent metals: the low temperature phases of Li and Na.

The expected absence of spin–orbit coupling leads to the use of a double or 'Jones' zone with an inscribed Fermi sphere shown in Fig. 7-10. (Glaunsinger $et\ al.$ 1972; Rosenthal and Maxfield 1973, 1974). There are two molecules of $Li(NH_3)_4$ per primitive cell. Here the Γ to A distance is $2 \cdot 9 \times 10^7\, cm^{-1}$ while $k_F = 5 \cdot 1 \times 10^7\, cm^{-1}$, thus there would be zone boundary contact over a large area were only the first zone considered. Towards the M-point the zone radius is $5 \cdot 2 \times 10^7\, cm^{-1}$ so contact there is unlikely unless the Fermi surface is not spherical. In the $double$ zone, the free-electron Fermi sphere does not make contact with the boundary. The proximity of the zone at the M-point (the $\langle 1100 \rangle$ direction), however, suggests that a minor distortion would lead to zone contact and in the absence of a gap a wide variety of orbits (hole and open orbits as well as electron orbits) becomes possible, as in copper.

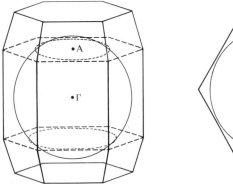
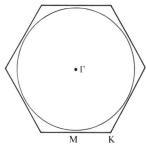

FIG. 7.10. Possible Fermi surface for hexagonal $Li(NH_3)_4$. (Rosenthal and Maxfield 1973, 1974).

Thompson and his co-workers (McDonald and Thompson 1966; Cate and Thompson 1971; LeMaster and Thompson 1972) have adopted the view that spin–orbit coupling may well open a gap at A, leading to *extensive* hole orbits and complete compensation such that $n_e = n_h$ where n_e and n_h are the densities of electrons and holes respectively. The mobilities need not be equal. Using this approximation one can make an internally consistent analysis of the known transport coefficients and even the magnetic susceptibility. One begins by writing the conductivity σ as

$$\sigma = n \, |e| \, (\mu_e + \mu_h) \tag{7.2}$$

where $n = n_e = n_h$ is the (common) value of electron and hole densities, and μ_e and μ_h are the respective mobilities. The Hall coefficient R_H is

$$R_H = (nec)^{-1}(\mu_e - \mu_h)/(\mu_e + \mu_h) \tag{7.3}$$

(Compare eqn 2.22) and the magnetoresistance in a field B is

$$\Delta\rho/\rho = AB^2 = \mu_e\mu_h B^2 \tag{7.4}$$

where $\rho = \sigma^{-1}$ and $\Delta\rho = \rho(B) - \rho$. The thermoelectric power due to diffusion only in the same approximation is:

$$S = [(\pi^2 k_B^2 T/3eE_F)(\mu_e - \mu_h)/(\mu_e + \mu_h), \tag{7.5}$$

where E_F is the Fermi energy appropriate to the density of either carrier and k_B is the Boltzmann factor. The susceptibility requires Fermi integrals at this low electron density (Glaunsinger *et al.* 1972), but with low densities (and therefore low Fermi energies) there is an appreciable temperature dependence to the Pauli susceptibility.

Considering only eqns (7.2) and (7.4) where $\sigma = 3\cdot2 \times 10^6 \, \Omega^{-1} \, cm^{-1}$ and $A = 46 \times 10^8 \, (cm^2 \, V^{-1} \, s^{-1})^2$ at 4·2 K (McDonald and Thompson 1966) one obtains an upper limit for n which is $1\cdot5 \times 10^{20} \, cm^{-3}$. Then adding eqn (7.3) with $R_H \approx 1\cdot4 \times 10^{-3} \, cm^3 \, c^{-1}$ the two mobilities may be extracted and are $\mu_e = 7 \times 10^4 \, cm^2 \, V^{-1} \, s^{-1}$ while $\mu_h = 6\cdot5 \times 10^4 \, cm^2 \, V^{-1} \, s^{-1}$ for $n = 1\cdot1 \times 10^{20} \, cm^{-3}$. Using these parameters and eqn (7.5) one obtains a thermopower of the form

$$S = (-0\cdot009 \, \mu V \, K^{-2})T$$

while the data give

$$S = 2\cdot6 \, \mu V \, K^{-1} - (0\cdot11 \, \mu V \, K^{-2})T$$

The diffusion thermopower does not seem to be adequate to recapture the observations. The susceptibility yields 3×10^{19} cm^{-3} for an effective mass ratio $m^*/m = 3\cdot0$. This value of n, together with the two resistivity eqns (7.2) and (7.4) leads to mobilities near 70 and $0\cdot7 \times 10^{-4}$ cm^2 V^{-1} s^{-1}. With such unequal mobilities, R_H is expected to be close to the free electron value, certainly within the error bars of Fig. 7.7. S is then computed to have a temperature coefficient of $-1\cdot9$ μV K^{-2} as far above the observation as the previous estimate was below.

There are two fundamental problems with the two-band model. Internally it fails to fit S and gives n temperature dependent at low temperatures. More important is the absence of a real justification for the spin–orbit splitting at the single zone top and bottom (point A, Fig. 7.10). Numbers below 10^{-3} eV are common and are just too small to produce a separation of states in the first and second zones. Only if there are appreciable errors in the X-ray results (or if a transition occurs in the 65–70 K range) can the two-band model be seriously considered.

The push to the compensated model came from the strong, non-saturating magnetoresistive effects. An alternative explanation can be found in terms of the open orbits allowed in the double (Jones) zone scheme based on zone boundary contact at M. If there are open orbits in the plane normal to the field then even a B^2 dependence can be obtained as assumed in eqn (7.4). In polycrystalline samples the exponent in the magnetoresistance decreases towards but not to unity while R_H acquires a field dependence. The former effect is consistent with the B$^{1\cdot5}$ power law observed in some cases but no field dependence has been reported for R_H. Rosenthal and Maxfield (1973, 1974) nevertheless conclude the open orbit effects due to contact near M may well explain the observations, particularly if there is extensive contact so that many open-orbit carriers exist.

If only a single band exists, as just concluded, then the T^2 term in the low temperature resistivity is difficult to rationalize. A large density of states and electron—electron scattering seems less likely than when originally proposed (Cate and Thompson 1971; David *et al.* 1973). While the low electron density ($r_s = 3\cdot8$Å, compared to $3\cdot0$ Å for Cs) suggests the possibility of departures from an independent particle model (Cohen and Thompson 1968) there are many other possible explanations for the observed effects including even antiferromagnetism (Glaun-singer *et al*, 1972).

If one could but make single crystals of the low tempera-ture phase of $Li(NH_3)_4$ most if not all of these questions could be resolved. Until then it appears that ambiguities and inconsisten-cies must persist.

7.3. Alkaline earth compounds

These materials have unknown melting points and have been prepared only by cooling metal solutions which contain excess NH_3. The eutectics lie near 8 m.p.m. (Schroeder, Thompson, and Oertel 1969) so that appreciable difficulties are encountered in attaining the exact 6:1 ratio. As in all the M–NH₃ solutions reactivity is also a problem and enclosing, protective atmospheres are required.

7.3.1. Structure of alkaline earth hexammines

The structure of the alkaline earth hexammines has been established by two independent investigations (Holland and Cagle 1963; Mammano and Sienko 1970) at 233 and 77 K. The structures and lattice parameters are collected in Table 7.2 to-gether with those of the other metal ammines. The nitrogens of the NH_3 molecules are arranged octahedrally around each metal atom. The Ca–N distance was found to be near 2·8 Å. There are two metal atoms per unit cell with metal–metal distances of 7·9, 8·3, and 8·6 Å in the Ca, Sr, and Ba. Low temperature data show

TABLE 7.2
Structure of metalammines

Formula	Temperature (K)	Lattice	Lattice constants (Å)	Density g/cm³
$Li(NH_3)_4$	77–82	hcp	$a = 7·0$	$0·57 \pm 0·03$
			$c = 11·1$	0·53
	82–88	bcc	9·55	
$Ca(NH_3)_6$	233	bcc	9·12	0·62
$Sr(NH_3)_6$	233	bcc	9·57	0·72
$Ba(NH_3)_6$	233	bcc	9·97	0·80
$Eu(NH_3)_6$	198	bcc	9·55	
$Yb(NH_3)_6$	198	bcc	9·30	

the expected decrease in lattice constant except for $Ca(NH_3)_6$. Since each material is likely to show significant deviations (Marshall and Hunt 1956) from stoichiometry, no particular significance can be attributed to any of the temperature variations.

7.3.2. Energetics of alkaline earth hexammines

The energies of formation of the metal hexammines computed following the scheme of Section 7.2.3 plus second ionization energies of these divalent metals are shown in Table 7.1. Senozan and coworkers (Dickman, Senozan, and Hunt 1970; Plummer and Senozan 1971; Mast and Senozan 1970; Frisbee and Senozan 1972; Senozan 1973) have determined the standard heats of dissociation at 273 K from a measurement of the NH_3 vapor pressure above the solid and they are tabulated in Table 7.3 with other thermodynamic parameters derivable from the vapor pressure; the ΔS values are calculated at temperatures such that $P = 1 \, \text{atm}^{-1}$. The ΔH values in Table 7.3 correspond to the energies associated with steps (b)–(g) of Section 7.2.3 since only the NH_3 is converted to the gas phase. That is, the ΔH of table 7.3 measures the energy in the reaction

$$M(s) + 6NH_3(g) \rightarrow M(NH_3)_6(s).$$

These energies might then be supposed to be equivalent to those estimated by Oesterreicher *et al.* excluding step (a). That the numbers do not match should be no great surprise considering the approximations made in estimating the various contributions. What is significant is that the trends observed by Senozan are consistent with expectations based on density (as it influences the

TABLE 7.3

Standard dissociation enthalpy, free energy and entropy of metalammines[a]

Compound	ΔH° (kcal/mole NH_3)	$\Delta F^{\circ \, b}$ (kcal/mole NH_3)	$\Delta S^{\circ c}$ (cal/deg mole NH_3)
$Ca(NH_3)_6$	$9 \cdot 95 \pm 0 \cdot 10$	$1 \cdot 89 \pm 0 \cdot 01$	$29 \cdot 5 \pm 0 \cdot 4$
$Sr(NH_3)_6$	$9 \cdot 74 \pm 0 \cdot 15$	$1 \cdot 68 \pm 0 \cdot 02$	$29 \cdot 5 \pm 0 \cdot 6$
$Ba(NH_3)_6$	$9 \cdot 25 \pm 0 \cdot 35$	$1 \cdot 22 \pm 0 \cdot 10$	$29 \cdot 5 \pm 1 \cdot 5$
$Eu(NH_3)_6$	$9 \cdot 46 \pm 0 \cdot 10$	$1 \cdot 11 \pm 0 \cdot 02$	$30 \cdot 6 \pm 0 \cdot 4$
$Yb(NH_3)_6$	$8 \cdot 92 \pm 0 \cdot 15$	$0 \cdot 62 \pm 0 \cdot 03$	$30 \cdot 4 \pm 0 \cdot 6$

[a] Frisbee and Senozan (1972). [b] at 0 °C. [c] at 1 atm.

electronic delocalization energy) and on ionic radius. The latter enters through the ammoniation energy (step d).

7.3.3. *Transport coefficients of alkaline earth hexammines*

Qualitative observations of the course of electrical resistance with temperature were made by Birch and MacDonald (1947, 1948) as they searched for superconductivity (Ogg 1946a). They mapped out portions of the phase diagram but reported nothing else. McDonald, Thompson, and Bowen (1964) reported the $Ca(NH_3)_6$ resistivity shown in Fig. 7.11. The measurements were made electrodelessly as described by McDonald and Thompson (1966). The limiting resistivity is 3 $\mu\Omega$ cm and there are signs of a solid–solid phase transition near 40 K.

7.3.4. *Magnetic data of alkaline earth hexammines*

Susceptibility measurements have been reported from the Cornell group (Oesterreicher *et al.* 1969; David *et al.* 1973) and are shown in Fig. 7.12. The behaviour is complex and indications of several solid–solid or magnetic phase transitions are present. In each case there are complications because of the lack of stoichiometry, though corrections have been attempted for the excess components. In all the magnetic work from Sienko's

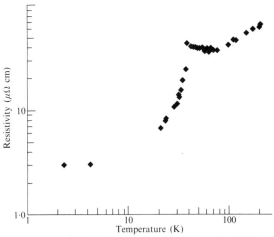

FIG. 7.11. Temperature dependence of resistivity of $Ca(NH_3)_6$. (McDonald, Thompson, and Bowen 1964).

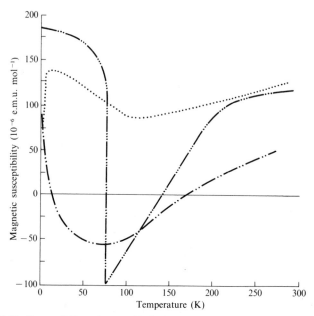

Fᴵɢ. 7.12. Susceptibility of several alkaline earth hexamines. See Fig. 2.1 for notation (David *et al.* 1973).

laboratory corrections for the effect of NH_3 at low temperatures (4 K) are particularly troublesome as their data for the pure NH_3 show an unexpected and unexplained rise as temperature is lowered in this range. It is not impossible that some impurity effects in the NH_3 has influenced all of their results. Such an effect cannot, however, be the source of the sign changes observed in $Sr(NH_3)_6$ and $Ba(NH_3)_6$. Future work will be required to provide sufficient data for corroboration and meaningful interpretation of these very interesting results.

7.4. Lanthanide–metal compounds

In both europium and ytterbium hexammines there is every evidence that the lanthanides go in as divalent ions (See Chapter 6) so that they might as well have been discussed with the alkaline earths. Indeed their structures and the energetics of their formation are little different and have already been included in Tables 7.1–7.3. Nothing further needs to be said on that point. However, the Eu^{2+} ion has a magnetic moment and the magnetic properties of $Eu(NH_3)_6$ are unique and deserving of separate discussion. No other data seem to have been reported.

7.4.1. *Magnetic properties of lanthanide–metal compounds*

Oesterreicher *et al.* (1969) have reported static susceptibilities while Brown, Cohen, and West (1973) and also Parker and Kaplan (1973*a*, *b*; 1974*a*, *b*) have used Mössbauer spectroscopy to determine the magnetic state of the Eu^{2+} ion.

There are two extraordinary features in the static susceptibility data for $Eu(NH_3)_6$ as may be seen in Fig. 7.13 (Oesterreicher *et al.* 1969). One is the decided increase in magnetic moment below 47 K and the other is the small but significant deviation from Curie–Weiss behaviour below an apparent Curie temperature of 5·5 K. The low temperature susceptibility is field dependent and Oesterreicher *et al.* (1969) suggest that ferromagnetism is setting in. They believe that these somewhat surprising results can be

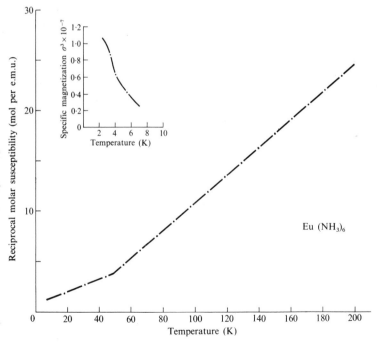

FIG. 7.13. Reciprocal molar susceptibility of $Eu(NH_3)_6$. The inset shows the low-temperature magnetization (Oesterreicher *et al.* 1969).

best explained by presuming the conduction electrons to be progressively frozen out into localized d-states as the temperature is lowered below 47 K. $Eu(NH_3)_6$ is then perhaps an example of an excitonic insulator (David *et al.* 1973). Ferromagnetic impurities, including the $Eu(NH_2)_2$ formed by the standard decomposition in the liquid (Section 1.5), may also play a role.

The replacement of Eu^{2+} by the nonmagnetic Yb^{2+} or Sr^{2+} ions does not appear to affect the Eu moment until only 10 per cent of the metal ions are Eu. There is little change in the Curie temperature ($\sim$6 K). Fig. 7.14 collects the data (Oesterreicher *et al.* 1969).

Parker and Kaplan (1973) find no evidence of carriers entering localized states in $Eu(NH_3)_6$ and also dispute the existence of magnetic ordering above their lowest data at 1·2 K. They attribute magnetic relaxation effects in the 4 to 70 K range to an Eu^{2+} dipole–dipole relaxation mechanism. The area of the Mössbauer resonance yields a Debye temperature of 43 K (compare 55 K in $Li(NH_3)_4$).

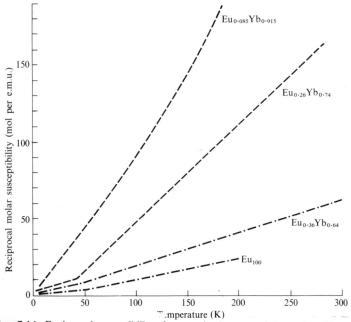

FIG. 7.14. Reciprocal susceptibility of some mixed rare-earth hexamines (Oesterreicher *et al.* 1969).

Each of these effects is consistent with the large metal–metal distance (~8 Å) characteristic of these and other metal-ammines and does not require so esoteric an explanation as an excitonic insulator. In any event more work is required.

The effects of further dilution of the Eu^{2+} magnetic moments, obtained by substituting Yb for Eu, do not provide any resolution of the discrepancy between Mössbauer and static susceptibility results. The latter experiments (Oesterreicher et al. 1969) failed to show any shift in the 5·5 K Curie point. It will be interesting to see the results of similar Mössbauer experiments.

7.5. Summary and discussion

The results reported in this chapter have opened up more questions than they have resolved. In the case of $Li(NH_3)_4$ the considerable available data is rendered nearly useless by the absence of single crystals. The hexammines can be studied in single crystal form, albeit with defects, but there is as yet insufficient data to yield a clear picture of what is going on. The effects of dilution of the electron gas and of magnetic moments (in the case of Eu) are central to any understanding of these materials, and should provide sufficient motivation to guarantee their eventual mastery.

One can also ask what about the other alkali metals. Why is there no $Na(NH_3)_x$? Attempts to produce such metallic solids have failed despite an occasional report of 'blue' solids formed from *dilute* solutions. (Catterall 1970; Dmitrenko and Shchetkin 1973).

It is probable that it is the ammoniation process (step d) that forms the bottleneck which prevents the formation of solid Na, K, etc. ammines. If one adopts, for example, the formula $Na(NH_3)_4$ suggested by the liquid diffusion data of Garroway and Cotts (1973), then the ion is sufficiently large for the transfer of lone pair electrons to sp^3 hybrid orbitals based on the Na to be energetically unfavourable and the solid does not exist despite comparable ion-dipole interactions. This is consistent with the occurrence of saturation in the liquid prior to the attainment of the $Na(NH_3)_4$ stoichiometry. Similar arguments should apply to K–NH_3 solids. Senozan (private communication) has suggested substituting Na for Ca may clarify the situation.

Cs–NH$_3$ solids are another problem. Two reports of a bronze solid exist (Birch and MacDonald 1948; Schroeder, Thompson, and Oertel 1969) whereas Lelieur and Rigny (1973) find only metallic Cs n.m.r. in the solid and totally reject the existence of a compound. One can easily imagine that steric hinderances would preclude solid compound formation, even if hybridization were possible with assistance from d-band electrons.

REFERENCES

BIRCH, A. J. and MacDONALD, D. K. C. (1947). *Trans. Faraday Soc.* **44**, 735.
—— (1948). *Oxf. Sci.* 1948, 1.
BOORSE, H. A., COOK, D. G., PONTIUS, R. B., and ZEMANSKY, M. W. (1946). *Phys. Rev.* **70**, 92.
BROWN, J. P., COHEN, R. L., and WEST, K. W. (1973). *Chem. Phys. Letts.* **20**, 271.
CATE, R. C. and THOMPSON, J. C. (1971). *Physics Chem. Solids* **32**, 443.
CATTERALL, R. (1970). *Phil. Mag.* **22**, 779.
COHEN, M. H. and THOMPSON, J. C. (1968). *Adv. Phys.* **17**, 857.
DAVID, T., GLAUNSINGER, W., ZOLOTOV, S., and SIENKO, M. J. (1973). In *Electrons in fluids* (eds J. Jortner and N. R. Kestner), p. 323. Springer-Verlag, Berlin, Heidelberg, New York.
DICKMAN, S., SENOZAN, N. M., and HUNT, R. L. (1970). *J. chem. Phys.* **52**, 2657.
DMITRENKO, I. M. and SHCHETKIN, I. S. (1973). *Zn. eksp. teor. Fiz. SSSR, Pis'ma Redakt.* **18**, 497.
DYE, J. L., CERASO, J. M., LOK, M. T., BARNETT, B. L., and TEHAN, F. J. (1974). *J. Am. chem. Soc.* **96**, 608.
FRISBEE, R. H. and SENOZAN, N. M. (1972). *J. chem. Phys.* **57**, 1248.
GARROWAY, A. N. and COTTS, R. M. (1973). *Phys. Rev.* A **7**, 635.
GLAUNSINGER, W. S., ZOLOTOV, S., and SIENKO, M. J. (1972). *J. chem. Phys.* **56**, 4756.
HARRISON, W. A. (1970). *Solid state theory.* McGraw-Hill, New York.
HERRING, C. (1960). *J. appl. Phys.* **31**, 1939.
HOLLAND, H. J. and CAGLE, F. W. (1963). 145th National Meeting of the American Chemical Society. (Unpublished).
JAFFE, H. (1935). *Z. Phys.* **93**, 741.
JOLLY, W. L. (1959). *Prog. inorg. Chem.* **1**, 235.
KLEINMAN, L., HYDER, S. B., THOMPSON, C. M., and THOMPSON, J. C. (1970). In *Metal–ammonia solutions* (eds J. J. Lagowski and M. J. Sienko), p. 229. Butterworths, London.
LELIEUR, J. P. and RIGNY, P. (1973). *J. chem. Phys.* **59**, 1148.
LEMASTER, E. W. and THOMPSON, J. C. (1972). *J. Solid State Chem.* **4**, 163.
LEVY, R. A. (1956). *Phys. Rev.* **102**, 31.
LO, R. E. (1966). *Z. anorg. Allg. Chem.* **344**, 230.
MAMMANO, N. (1970). In *Metal–ammonia solutions* (eds J. J. Lagowski and M. J. Sienko), p. 367, Butterworths, London.
—— and COULTER, L. V. (1967). *J. chem. Phys.* **47**, 1564.
—— —— (1969). *J. chem. Phys.* **50**, 393.
—— and SIENKO, M. J. (1968). *J. Am. chem. Soc.* **90**, 6322.
—— —— (1970). *J. Solid State Chem.* **1**, 534.
MARSHALL, P. (1964). In *Metal–ammonia solutions* (eds G. Lepoutre and M. J. Sienko), p. 97. Benjamin, New York.

MARSHALL, P., and HUNT, H. (1956). *J. phys. Chem.* **60**, 732.
MAST, G. and SENOZAN, N. M. (1970). *J. chem. Phys.* **53**, 1296.
MCDONALD, W. J. and THOMPSON, J. C. (1966). *Phys. Rev.* **150**, 602.
—— —— and BOWEN, D. E. (1964). *Bull. Am. chem. Soc.* **9**, 735,
MORGAN, J. A., SCHROEDER, R. L., and THOMPSON J. C. (1965). *J. chem. Phys.* **43**, 4494.
—— and THOMPSON, J. C. (1967). *J. chem. Phys.* **47**, 4607.
MOTT, N. F. and JONES, H. (1936). *The theory of the properties of metals and alloys.* Clarendon Press, Oxford.
MUELLER, W. E. (1969). *Appl. Optics.* **8**, 2083.
—— and THOMPSON, J. C. (1969). *Phys. Rev. Letts.* **23**, 1037.
OESTEREICHER, H., MAMMANO, N., and SIENKO, M. J. (1969). *J. Solid State Chem.* **1**, 10.
OGG, R. A., JR. (1946a). *Phys. Rev.* **69**, 243.
—— (1946b). *Phys. Rev.* **70**, 93.
PARKER, J. T. and KAPLAN, M. (1973a). *Phys. Rev.* B **8**, 4318.
—— —— (1973b). *Chem. Phys. Letts.* **23**, 437.
—— —— (1974a). *Chem. Phys. Lett.* **24**, 280.
—— —— (1974b). *J. chem. Phys.* **60**, 1328.
PLUMMER G. and SENOZAN, N. M. (1971). *J. chem. Phys.* **55**, 4062.
ROSENTHAL, M. D. and MAXFIELD, B. W. (1973). *J. Solid State Chem.* **7**, 109.
—— —— (1974). *J. Low Temp. Phys.* **14**, 15.
SCHROEDER, R. L., THOMPSON, J. C., and OERTEL, P. L. (1969). *Phys. Rev.* **178**, 298.
SENOZAN, N. M. (1973). *J. inorg. nucl. Chem.* **35**, 727.
SIENKO, M. J. (1964). In *Metal–ammonia solutions* (eds G. Lepoutre and M. J. Sienko), p. 23. Benjamin, New York.
VARLASHKIN, P. G. and ARIAS-LIMONTA, J. A. (1971). *J. chem. Phys.* **54**, 1230.
ZIMAN, J. M. (1964). *The principles of the theory of solids.* Cambridge University Press.

8

SOLUTIONS IN OTHER SOLVENTS

8.1. Introduction

THERE are a number of solvents other than NH_3 which are known to dissolve metals in significant quantities. These solvents share with ammonia (a) a kinetic inertness to the acceptance of an extra electron, (b) effective cation solvation (Sundheim 1959), and (c) ready purification (Dye 1973). In addition to the solvents listed in Table 8.1 and their mixtures there are a number of other solvents, which were once thought to dissolve minute amounts of metals, e.g. 10^{-4} m.p.m. More recent research (Dalton, Ryjnbrandt, Hansen, and Dye 1966) suggests, however, that trace amounts of the solvents in Table 8.1 or of NH_3 were responsible for the observed solubility. Reactions such as ethylamine → ethylene + ammonia, catalyzed by metal, might well be the source of NH_3 in apparently pure solvents, though this has been disputed by Hurley, Tuttle, and Golden (1970).

It is also possible to increase the solubility limits in such materials by the addition of certain cyclic polyethers: 'crown' or 'cryptate' (Dye 1973). While there is some debate as to whether or not the additives *only* tie up the metal cation or have influence on the electron as well (Dorfman and Jou 1973) there is no doubt that solubilities go up.

None of the metals may be dissolved in sufficient quantities to yield free electrons instead of solvated or otherwise localized electrons. Reports of bronze phases in mixed solvents (Dalton, Ryjnbrandt, Hansen, and Dye 1966) containing large amounts of NH_3 probably represent phase separation of the metal-containing NH_3 from the other solvent. HMPA† solutions characterized (Catterall, Stodulski, and Symons 1970) as 'bronze' have the reddish colour of tungsten bronzes rather than the golden metallic lustre of concentrated M–NH_3 solutions. The usual precautions required for successful M–NH_3 experiments are not to be ignored here.

† See Table 8.1 for full name of this and other solvents.

TABLE 8.1

Solutions showing solvated electron and M^- spectra[a,b]

Formula	Name	Short name(s)	ε_0 near room temperature	Solvated electron			Metal anion		
				$\hbar\omega_{max}$ (eV)	Half-width (eV)	$-d(\hbar\omega_{max})/dT$ (10^{-3} eV/K)	Species	$\hbar\omega_{max}$ (eV)	$d\hbar\omega_{max}/dT$ (10^{-3} eV/K)
$CH_2(OH)CH$-$(OH)CH_2OH$	glycerol[c]	—	42.5	2.35	1.5	—	—	—	—
$CH_2(OH)CH_2OH$	ethylene glycol[c]	EG	38.7	2.16	1.35	—	—	—	—
CH_3OH	methanol[c]	MeOH	32.6	1.97	1.29	—	—	—	—
CH_3CH_2OH	ethanol[c]	EtOH	24.5	1.77	1.55	—	—	—	—
H_2O	water	H_2O	78.5	1.72	0.92	2.8	—	—	—
$CH_3CH(OH)CH_3$	isopropanol[c]	iPrOH	18.6	1.51	1.22	1.22	—	—	—
CH_3NH_2	methylamine[d]	$MeNH_2$; MA	9.4	0.95	—	—	—	—	—
$(CH_2)_2(NH_2)_2$	ethylenediamine	EDA	12.9	0.94	0.88	1.9	Na^-	1.52	1.71
NH_3	ammonia	NH_3	15.9	0.8	0.4	1.5	—	—	—
$CH_3CH_2NH_2$	ethylamine	$EtNH_2$; EA	6.3	0.69	—	2.9	Na^-	1.75	2.02
$(CH_3CH_2)_2NH$	diethylamine	DEA	5.5	0.65	—	—	Na^-	1.71	2.49
$CH_3OCH_2CH_2$-$OCH_2CH_2OCH_3$	bis(2-methoxy-ethyl)ether	diglyme	—	0.65	0.58	—	Na^-	1.73	2.33
$(CH_3CH_2)_2O$	diethyl ether	DEE	4.3	0.64	0.43	1.5	Na^-	1.61	2.12
$CH_3OCH_2CH_2OCH_3$	dimethoxyethane	DME	7.2	0.60	0.50	—	Na^-	1.71	2.09
$OCH_2CH_2CH_2CH_2$	tetrahydrofuran	THF	7.4	0.58	0.43	—	Na^-	1.69	2.06
$OCH_2(CH_3)$-$CH_2CH_2CH_2$	methyltetra-hydrofuran	MTHF	—	0.57	0.45	—	—	—	—
$[(CH_3)_2N]_3PO$	hexamethylphos-phorictriamide	HMPA	30	0.55	—	—	Na^-	1.62	—

[a] Dye (1973). [b] Dorfman and Jou (1973). [c] Arai and Sauer (1966). [d] Hohlstein and Wannagat (1956).

There are many effects associated with specific cations (Section 8.21), in contrast to NH_3 solutions where the cation has little influence. Since Na may be leached from Pyrex glass vessels (Hurley, Tuttle, and Golden 1970) there has been a long history of confusion over which cation does what. That confusion has recently been resolved but older papers must be read carefully for Na^+ effects as well as for impure solvent effects.

The solvated electron may be studied in a larger class of fluids than those in Table 8.1, even H_2O, by using pulse radiolysis (Dorfman 1973) and similar techniques (Section 3.4.4).

This chapter, then, will parallel Chapter 3 with broad subdivision as to whether stable solutions of metals can be formed or if electrons must be introduced by other means. Topics such as those of Chapters 2, 4, and 5 do not appear, save in the last section (Section 8.6) where some evidence of an incomplete M–NM transition is presented. Recent reviews have been given by a number of authors (Hart 1964; Lagowski and Sienko 1970; Dye 1973) and reduce the coverage required here.

8.2. Metal solutions in amines

Methylamine, ethylamine, and ethylenediamine have been more extensively studied than most of the other solvents; $MeNH_2$ dissolves more metal than any other (Nakamura, Horie, and Shimoji 1974), though there are severe problems with NH_3 contamination that render much data questionable. Since optical data are most abundant and most informative, they will be presented first, then any other available data.

8.2.1. *Optical data*

There are two absorption peaks (Lok, Tehan, and Dye 1972), one near 0·7 eV, which is independent of solute, and another in the range 1·5–1·8 eV, which is strongly dependent on the solute (Matalon, Golden, and Ottolenghi 1969) as well as the solvent. Typical spectra are shown in Fig. 8.1 for THF (Lok, Tehan, and Dye 1972). Similar results are obtained for amines. The solute-independent peak is essentially the same as that obtained for injected electrons (Dalton, Dye, Fielden, and Hart 1966), is attributed to the solvated electron (Section 8.4), and will be discussed in that section. The solute-dependent peak position has

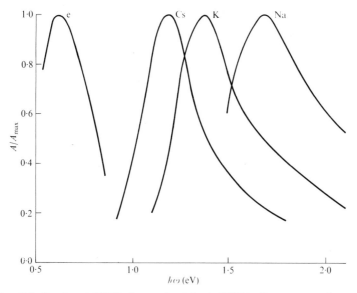

FIG. 8.1. Spectra at 298 K of several species in THF in the presence of crown cryptate (Dye 1973).

been tabulated in Table 8.1 along with other data on the amines and will be discussed in the present section.

The characteristics of the solute-dependent peak are as follows:

1. strong dependence of position on solute;
2. negative values of $d(\hbar\omega_{max})/dT$ of order 10^{-3} eV K^{-1};
3. shifts of position and temperature dependence are correlated. Fig. 8.2 shows the peak position as a function of temperature coefficient for Na in a number of solvents;
4. shift of the M$^-$ (Section 8.2.5) peak to lower energies for larger *anions*, for a given cation;
5. absorption bands narrow at lower temperature;
6. a comparison of peak positions for Na and K in the various solvents reveals correlated shifts so that whatever shifts the band produced by dissolving Na has the same effect on the band produced by K.

Debacker and Dye (1971) have determined the extinction coefficient of the metal-dependent absorption for Na in EDA. This yields an oscillator strength of 1·9, upon assuming one sodium nucleus per absorbing unit. From the f-sum rule such an

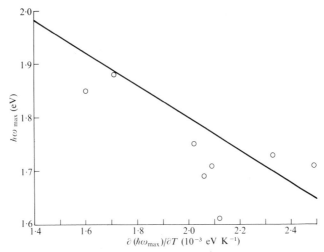

FIG. 8.2. Correlation between Na⁻ peak position at 298 K and its temperature coefficient. The slope of the straight line is 298 K (Lok, Tehan, and Dye 1972).

oscillator strength would require that *two* electrons be involved in the species undergoing the transition.

These observations will be attributed (Section 8.2.5) to the existence of species of stoichiometry M^-.

8.2.2. *Electron transport*

Conductance data have been obtained for both Li– and Na–MeNH$_2$ over the entire concentration range (Evers, Young, and Pauson 1957; Berns, Evers, and Frank 1960; Evers and Longo 1966; Dewald and Browall 1970; Nakamura, Yamamoto, and Shimoji 1973). In addition there are data for dilute solutions of several metals in EDA (Dewald and Dye 1964). The dilute solutions are all much alike when viscosity is taken into account as may be seen in Fig. 8.3 where the product of conductivity and *solvent* viscosity is ordinate. The data for more concentrated solutions is compared to M–NH$_3$ data in Fig. 8.4; the amine solutions have a two order-of magnitude lower conductivity even near saturation. Mott's minimum conductivity (Section 4.5.5) is never attained.

The temperature coefficient of conductivity is shown in Fig. 8.5 along with Li–NH$_3$ data similar to Fig. 2.5. The MeNH$_2$ results lie well above the NH$_3$ data over most of the range.

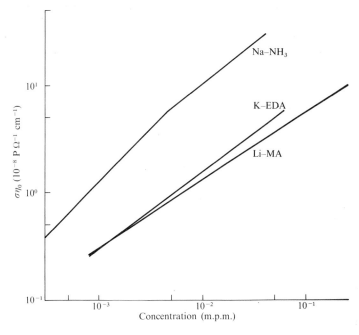

FIG. 8.3. Walden product (conductivity × solvent viscosity) for several solutions
(Dewald and Dye 1964; Dewald and Browall 1970).

8.2.3. *Magnetic resonance and static susceptibility*

The e.s.r. spectra obtained in amines are quite complicated, in
contrast to the single narrow line observed in NH_3 solutions
(Section 3.3.2) and in HMPA solutions (Catterall, Stodulski, and
Symons 1968). There is, in some cases, a single line attributable
to the solvated electron. But in most cases a multiline spectrum is
found caused by splitting from a single nucleus (Bar-Eli and
Tuttle 1964). The hyperfine splitting (h.f.s.) constant increases
with increasing temperature while the spectroscopic g-factor is
either constant or decreases with increasing temperature (see
Nicely and Dye (1970) for a compilation of references). The
linewidth increases with increasing temperature. Many of these
observations may be influenced by decomposition, by Na leached
from the glass, and by solvent impurities.

For Li in ethylamine a composite spectrum is observed (Bar-
Eli and Tuttle 1964). It consists of a 9-line hyperfine pattern

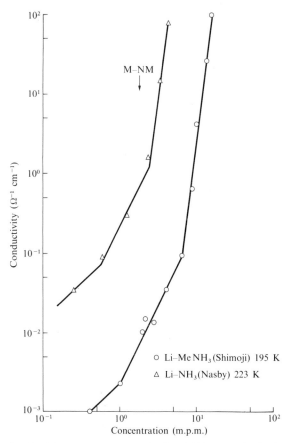

FIG. 8.4. Conductivity versus concentration in Li–NH$_3$ and Li–MeNH$_2$ solutions at 223 K and 195 K, respectively. The M–NM transition in Li–NH$_3$ is marked with an arrow (Nasby and Thompson 1970; Nakamura, *et al.* 1973).

superposed on a single relatively sharp line with the same *g*-value. This result has been confirmed by Catterall, Symons, and Tipping (1970).

The Li Knight shift in MeNH$_2$ (Haynes and Evers 1970) is essentially independent of concentration and a linearly decreasing function of inverse temperature. Values range from 15 p.p.m. at $+30\,°C$ to near 40 p.p.m. at $-70\,°C$. These values exceed those found in Li–NH$_3$ at all concentrations. It appears that the value of $|\psi(\text{Li})|^2$ (Section 2.8) in MeNH$_2$ exceeds that in NH$_3$ which in

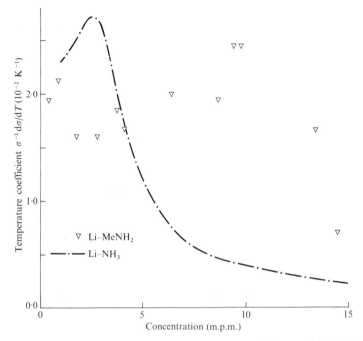

FIG. 8.5. Temperature coefficient of conductivity in Li–NH₃ and Li–MeNH₂ (Nasby and Thompson 1970; Evers and Longo 1966; Nakamura, *et al.* 1973).

turn exceeds that in $EtNH_2$ (see discussion following Haynes and Evers 1970). In each case the ratio is about ten.

The static susceptibility of Li–MeNH₂ is a slowly decreasing function of concentration in the range 2–10 m.p.m. (Nakamura, Yamamoto, and Shimoji 1973) and not far from the values expected for free spins.

8.2.4. *Other data*

Other measurements have been mostly limited to methylamine solutions and to Shimoji's laboratory. They include density (Longo 1970; Yamamoto, Nakamura, and Shimoji 1971), viscosity (Yamamoto, Nakamura, and Shimoji 1972) and vapour pressure (Nakamura, Horie, and Shimoji 1974).

The excess volume ΔV (eqn 3.8) is essentially independent of concentration and is in the range $40\text{–}50 \text{ cm}^3 \text{ mol}^{-1}$ (Yamamoto *et al.* 1971). The data reported by Longo (1970) were taken by

Evers and Harrison and show the kind of spurious behaviour at
low concentration which has already been discussed in Section
3.5.1 for M–NH₃ solutions. There is also an unexplained discrepancy of 25 per cent between the the two reports.

The viscosity η is an increasing function of concentration in
contrast to M–NH₃ data (Sections 3.5.3 and 3.8) and analogous
to salt data. In all cases η is a decreasing function of temperature.
Values lie in the range 0·3–0·6 centipoise, slightly above those
reported for M–NH₃ solutions.

The vapour pressures have been used to calculate chemical
potentials (activities, Section 3.13). The trends are quite close to
those observed in M–NH₃ solutions. In particular, the activity of
$MeNH_2$ departs positively from ideality near 0·08 m.p.m. and
falls well below the ideal curve above 0·12 m.p.m. These same
features, Fig. 2.20, are observed at 0·05 and 0·10 m.p.m. in
Na–NH₃ solutions. These features shift much less between the
two solvents than do the properties previously discussed.

8.2.5. *Models of metal–amine solutions*

Three species are generally assumed to exist in amine solutions
and may be described by the following reactions (Dye 1973):

$$2M \text{ (solid)} \rightarrow M^- + M^+ \tag{8.1}$$

$$M^- \rightarrow M + e^- \tag{8.2}$$

$$M \rightarrow M^+ + e^- \tag{8.3}$$

All species are solvated. Strongly solvating liquids and high
dielectric constants (see Table 8.1) tend to shift these equilibria
toward solvated cations and solvated electrons as in NH₃. Cation
solvation (Sundheim 1959) is important for releasing the electron
to the solvent.

The species M is generally regarded as a classic monomer in
the sense first proposed by Becker, Lindquist, and Alder (1956).
That is, the electron orbits on the *outside* of the first solvation
layer. Such a configuration is consistent with the h.f.s. observed in
the e.s.r. (Bar-Eli and Tuttle 1964; O'Reilly and Tsang 1965).
The species must live at least 10^{-6} s (Catterall, Symons, and
Tipping 1970) in the higher amines. However, just as ion pairs
are a minor species in NH₃, the concentration of M is never high

in amine solutions. Catterall, Symons, and Tipping (1970) prefer an ion-pair for M–EtNH$_2$ solutions, especially if M is Li.

As noted earlier, the metal-dependent optical absorption found in the visible part of the spectrum is attributed to a species with the stoichiometry M$^-$ (Matalon, Golden, and Ottolenghi 1969). Presumably this species is more like the monomer than the loosely coupled aggregates of solvated cations and electrons used in describing M–NH$_3$ solutions. However, there is strong evidence that the 2s electrons reside *within* the solvent shell rather than *without*. This conclusion is based in part on the fact that the M$^-$ n.m.r. chemical shift is independent of solvent (Ceraso and Dye 1974) though the idea has also been offered by Golden, Guttman, and Tuttle (1965) for M–NH$_3$ solutions.

The salt-like trends observed in the viscosity suggest that the solvated electron is not so important in metal solutions in amines as in NH$_3$.

8.3. Metal solutions in ethers and HMPA

Solubilities in ethers are generally quite low but may be increased significantly by the use of cyclic polyethers as cation complexing agents (Dye 1973). Nevertheless most data have been limited to e.s.r. and optical absorption.

As already shown in Fig. 8.1 there are optical features attributable to both the solvated electron and M$^-$ species (Section 8.2.5). The polyethers (crown and cryptate) may possibly perturb the solvated electron absorption as the lines are not identical to those obtained by radiolysis (Section 8.5). Concentration differences may also be important in this connection. Temperature effects on optical absorption are large, lying in the range 10^{-2} eV K^{-1}.

In all other respects the ethers and amines appear comparable.

8.4. Radiolysis

The techniques of radiolysis and photolysis described in Section 3.4.4 are applicable to many fluids and have been used to inject electrons into H$_2$O, NH$_3$, HMPA, the amines, and ethers.

Even hydrocarbons and liquid helium can be studied this way. Materials such as those of Table 8.1 yield solvated electrons with spectra essentially identical to those obtained by dissolving metals. Fig. 8.6 compares peak positions for the two sources of

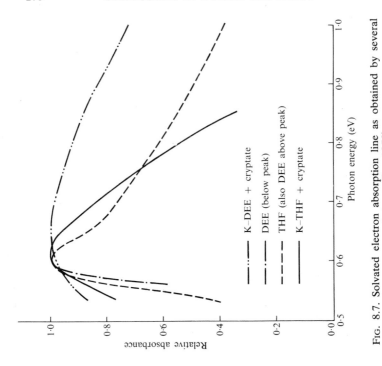

FIG. 8.7. Solvated electron absorption line as obtained by several techniques (Dye 1973).

K–DEE + cryptate

DEE (below peak)

THF (also DEE above peak)

K–THF + cryptate

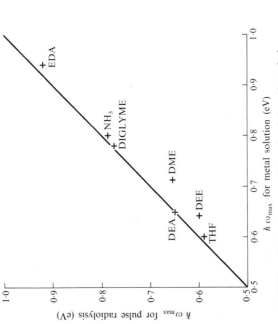

FIG. 8.6. The relation between the peak of the solvated electron absorption line obtained by dissolving metals and by radiolysis. The straight line has unit slope (Dye 1973).

solvated electrons. There is a red-shift of the spectrum of the radiolysis with respect to the metal by as much as 0·1 eV in some cases. The band is broader with metal and solubilizing agent present as may be seen in Fig. 8.7. Whether the differences arise from the use of cryptate, from cation concentration effects (even in radiolysis a small amount of salt is added for stability), from decomposition of solvent by metal, from spin pairing, or whatever, has not been established.

A compilation of the absorption spectra of solvated electrons in various solvents at room temperature is shown in Fig. 8.8 (Dorfman and Jou 1973; Jou and Dorfman 1973). Double arrows indicate peak locations when the band is not shown. One can separate out three broad classes from the peaks of Fig. 8.7. Those solvents containing OH⁻ groups lie at the highest energy, NH₃ and two of the amines lie in the middle, while the rest of the amines and the ethers have absorption peaks at low energies.

Data have also been obtained for several binary mixtures which show a single peak intermediate to those of the pure components with position dependent upon concentration and temperature

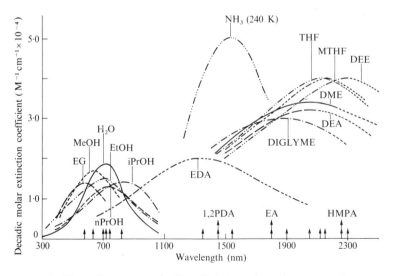

FIG. 8.8. Absorption spectra of solvated electrons in various pure solvents at room temperature. Peaks of spectra shown are indicated by arrows; other peaks by double arrows. See Table 8.1 for solvent names (Dorfman and Jou 1973).

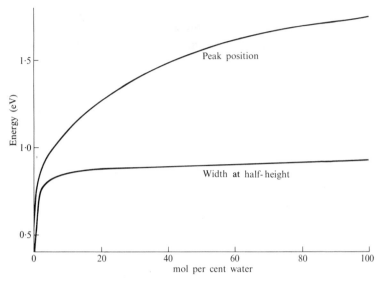

F$_{IG}$. 8.9. Absorption maximum and width at half-height of the solvated electron band in THF–H$_2$O mixtures at room temperature.

Fig. 8.9 shows how the peak position and halfwidth shift in an H$_2$O–THF mixture. Clearly the water dominates, though correlations with the parameters of Jortner's simple cavity model (Section 3.10.1) are not possible.

8.5. Solvated electron theory revisited

These data provide a fertile ground for testing the model of Section 3.10.2. Such tests have been carried out for only a very few of the materials for which spectra are shown in Fig. 8.7. The results are somewhat disappointing.

As already noted, correlations with the parameters of Jortner's simple model are generally not successful. When one stays within a single series, such as the alcohols, then success is somewhat more likely (Fueki, Feng, and Kevan 1972; Kestner 1973).

The solvated electron in H$_2$O has been studied extensively and there is good quantitative agreement though the predicted width and asymmetry of the absorption line do not match the observed values. As in NH$_3$ medium polarization effects are important for localization (Fueki, Feng, Kevan, and Christofferson 1971; Newton 1973; Kestner 1973). The void radius is in the range 1·0–1·4 Å, less than the size of an H_2O molecule and, of course, much less than in NH_3. The parameter r_d is 2·45 Å (see Fig. 3.37). Also as in NH_3 about 84 per cent of the electronic charge is within the first solvation layer though the tails of the wave function extend far into the medium. See Kestner (1973) for other references and comments.

8.6. A metal–nonmetal transition?

The conductivity data shown in Fig. 8.4 led Nakamura, Yamamoto, and Shimoji (1973; Nakamura, Horie, and Shimoji 1974) to suggest that a metal–nonmetal transition (Chapter 4) might occur in the $0·1 < x_{Li} < 0·2$ range in Li–MeNH₂ solutions. This supposition will now be examined. It is important to note that the nonmetallic methylamine solutions contain cation dependent species (Section 8.2.5) of sorts not presently expected in M–NH₃ solutions. Electron delocalization may well proceed in quite different ways in the two systems. There is no report of a bronze phase in the amines.

The metal–amine conductivities shown in Fig. 8.4 bear a superficial resemblance to the Li–NH₃ data. In particular, there is a two-order-of-magnitude increase of the conductivity between 7 and 11 m.p.m. in the Li–MeNH₂ solution; however, the increase is from 10^{-1} to $10^1 \, \Omega^{-1} \, cm^{-1}$. While these trends compare qualitatively to the 10^1 to $10^3 \, \Omega^{-1} \, cm^{-1}$ increase between 3 and 7 m.p.m. in Li–NH₃ solutions; the high conductivity state can hardly be called metallic. Fig. 8.10 shows also that the detailed course of σ with concentration is quite different even if the order of magnitude is suppressed.

Susceptibilities (Nakamura *et al.* 1973) and Knight shifts (Nakamura, Yamamoto, Shimokawa, and Shimoji 1971) are consistent with unpaired electron spins and fail to show the kind of transition characteristic of M–NH₃ solutions.

Models of the M–NM transition such as Mott's (Section 4.7.2) would require that the system be in the semiconducting regime with conduction proceeding by carriers activated to a mobility edge as in Li–NH₃ solutions below 3 m.p.m. Thermopower data would be quite interesting. The inhomogeneous model of Jortner

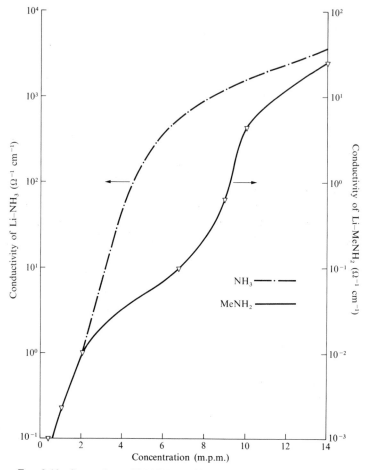

FIG. 8.10. Comparison of M–NM transition. Note the different ordinates.

and Cohen (Section 4.7.3) would require that the conductivities of the two phases be rather close. However, there is clear thermodynamic evidence (Nakamura, Horie, and Shimoji 1974) *against* the existence of fluctuations such as those found in Li–NH₃ solutions (Section 4.7.3). A different origin for inhomogeneities must be found if the Jortner–Cohen model is to be used.

The singularities in structure, as revealed by the chemical potential, together with the contrasting electronic properties suggest that the M–NM transition in M–NH$_3$ solutions is unrelated to the phase separation. It remains to be seen if a metallic phase can be induced in MeNH$_2$ solutions by pressure or by use of a crown or cryptate (Section 8.1).

A possible explanation for the absence of a metallic phase lies in the model for M$^-$ (Section 8.2.5). If the metal anion and its associated electrons lie within a shell of large methylamine molecules, as envisioned by Golden, Guttman, and Tuttle (1965, 1966) and corroborated by Ceraso and Dye (1974), then there must be very little overlap of electron wavefunctions so that the tight-binding band is quite narrow. It is therefore not possible to reach the condition for a metal–insulator transition, which relates the band-width to the intra-atomic interaction.

In addition, one may observe that in M–NH$_3$ solutions the localization of the electron on sites separate from the cation inevitably leads, at high concentrations, to a stronger tendency for the solvated electron wavefunction to have appreciable amplitude outside the cavity as it is attracted to the positive ion. Overlap is enhanced and the M–NM transition occurs. In MeNH$_2$, on the other hand, the larger size of the solvent molecule together with the tighter binding afforded by the presence of M$^+$ *within* the two-electron cavity (to use the language of the model appropriate to a M–NH$_3$ solution) weakens the attraction of solvated metal cations and apparently precludes a metallic phase.

Many more data will be required to establish this viewpoint.

8.7. Summary

Though solvated electrons can be produced in a wide variety of solvents both by electrolysis and by dissolving metals, there is little other similarity between such systems and M–NH$_3$ solutions. The generally lower dielectric constants and the generally larger molecules combine to reduce the interactions which are the origin of the rich range of physical properties observed in ammonia. While there is a wider variety of 'atomic' species, e.g. M$^-$, these species as well as the solvated electron remain isolated.

The solvated electron retains its unique properties in each solvent. Though details of size and coordination number may vary the basic structure of Jortner's model persists. So also the problems of line width and line shape persist.

REFERENCES

ARAI, S. and SAUER, M. C. (1966). *J. chem. Phys.* **44,** 2297.

BAR-ELI, K. and TUTTLE, T. R. (1964). *J. chem. Phys.* **40,** 2508.

BECKER, E., LINDQUIST, R. H., and ALDER, B. J. (1956). *J. chem. Phys.* **25,** 971.

BERNS, D. S., EVERS, E. C., and FRANK, P. W. (1960). *J. Am. chem. Soc.* **82,** 310; 1960.

CATTERALL, R., STODULSKI, L. P., and SYMONS, M. C. R. (1968). *J. chem. Soc. A,* **437.**

——, SYMONS, M. C. R., and TIPPING, J. W. (1970). In *Metal–ammonia solutions* (eds J. J. Lagowski and M. J. Sienko), p. 317. Butterworths, London.

CERASO, J. M. and DYE, J. L. (1974). *J. chem. Phys.* **61,** 1585.

DALTON, L. R., DYE, J. L., FIELDEN, E. M., and HART, E. J. (1966). *J. phys. Chem.* **70,** 3358.

——, RYNBRANDT, J. D., HANSON, L. M., and DYE, J. L. (1966). *J. chem. Phys.* **44,** 3969.

DEBACKER, M. G. and DYE, J. L. (1971). *J. phys. Chem.* **75,** 3092.

DEWALD, R. R. and BROWALL, K. W. (1970). *J. phys. Chem.* **74,** 129.

—— and DYE, J. L. (1964). *J. phys. Chem.* **68,** 128.

DORFMAN, L. M. (1973). In *Investigation of rates and mechanisms of reaction.* Part II: *Elementary reactions in solution and very fast reactions,* Chapter 11. Wiley, New York.

—— and JOU, F. Y. (1973). In *Electrons in fluids* (eds J. Jortner and N. R. Kestner), p. 447. Springer-Verlag, Heidelberg.

DYE, J. L. (1973). In *Electrons in fluids* (eds J. Jortner and N. R. Kestner), p. 77. Springer-Verlag, Heidelberg.

EVERS, E. C. and LONGO, F. (1966). *J. phys. Chem.* **70,** 426.

——, YOUNG, A. E., and PANSON, A. J. (1957). *J. Am. chem. Soc.* **79,** 5118.

FUEKI, F., FENG, D. F., and KEVAN, L. (1972). *J. chem. Phys.* **56,** 5351.

——, ——, ——, and CHRISTOFFERSON, R. (1971). *J. phys. Chem.* **75,** 2297.

GOLDEN, S., GUTTMAN, C., and TUTTLE, T. R. (1965). *J. Am. chem. Soc.* **87,** 135.

——, ——, —— (1966a). *J. chem. Phys.* **44,** 3791.

——, ——, —— (1966b). *J. chem. Phys.* **45,** 2206.

HART, E. J. (1964). *Science N.Y.* **146,** 19.

HAYNES, R. and EVERS, E. C. (1970). In *Metal–ammonia solutions* (eds J. J. Lagowski and M. J. Sienko), p. 159. Butterworths, London.

HOHLSTEIN, G. and WANNAGAT, U. (1956). *Z. Anorg. Allg. Chem.* **288,** 193.

HURLEY, I., TUTTLE, T. R., and GOLDEN, S. (1970). In *Metal–ammonia solutions* (eds J. J. Lagowski and M. J. Sienko), p. 449. Butterworths, London.

JOU, F. Y. and DORFMAN, L. M. (1973). *J. chem. Phys.* **58,** 4715.

KESTNER, N. R. (1973). In *Electrons in fluids* (eds J. Jortner and N. R. Kestner), p. 1. Springer-Verlag, Heidelberg.

LAGOWSKI, J. J. and SIENKO, M. J. (1970) (eds). *Metal–ammonia solutions.* Butterworths, London.

LOK, M. T., TEHAN, F. J., and DYE, J. L. (1972). *J. phys. Chem.* **76,** 2975.

LONGO, F. R. (1970). In *Metal–ammonia solutions* (eds J. J. Lagowski and M. J. Sienko), p. 493. Butterworths, London.

MATALON, S., GOLDEN, S., and OTTOLENGHI, M. (1969). *J. phys. Chem.* **73,** 3098.

NAKAMURA, Y., HORIE, Y., and SHIMOJI, M. (1974). *Discuss. Faraday Soc.* I. **70,** 1376.

——, YAMAMOTO, M., and SHIMOJI, M. (1973). In *Properties of liquid metals* (ed. S. Takeuchi), p. 385. Taylor and Francis, London.

——, ——, SHIMOKAWA, S., and SHIMOJI, M. (1971). *Bull. chem. Soc. Japan* **44,** 3212.

NASBY, R. D. and THOMPSON, J. C. (1970). *J. chem. Phys.* **53,** 109.
NEWTON, M. (1973). *J. chem. Phys.* **58,** 5833.
NICELY, V. A. and DYE, J. L. (1970). *J. chem. Phys.* **53,** 119.
O'REILLY, D. E. and TSANG, T. (1965). *J. chem. Phys.* **42,** 5333.
SUNDHEIM, B. R. (1959). *Trans. N.Y. Acad. Sci.* **21,** 281.
YAMAMOTO, M., NAKAMURA, Y., and SHIMOJI, M. (1971). *Trans. Faraday Soc.*
    67, 2292.
——, ——, —— (1972). *Discuss. Faraday Soc.* II **68,** 135.

AUTHOR INDEX

SUBJECT INDEX